AF247593

HYDROGEN: ITS TECHNOLOGY AND IMPLICATIONS

Editors

Kenneth E. Cox
Project Manager
Thermochemical Hydrogen
Los Alamos Scientific Laboratory
University of California
Los Alamos, New Mexico

K. D. Williamson, Jr.
Assistant Division Leader
Los Alamos Scientific Laboratory
University of California
Los Alamos, New Mexico

SERIES OUTLINE

Volume I
Hydrogen Production Technology

Volume II
Transmission and Storage of Hydrogen

Volume III
Hydrogen Properties

Volume IV
Utilization of Hydrogen

Volume V
Implications of Hydrogen Energy

Hydrogen: Its Technology and Implications

Volume IV
Utilization of Hydrogen

Editors:

Kenneth E. Cox

Project Manager
Thermochemical Hydrogen Project
Los Alamos Scientific Laboratory
University of California
Los Alamos, New Mexico

K. D. Williamson, Jr.

Associate Group Leader
Los Alamos Scientific Laboratory
University of California
Los Alamos, New Mexico

CRC PRESS, Inc.
Boca Raton, Florida 33431

Library of Congress Cataloging in Publication Data

Main entry under title:

Utilization of hydrogen.

   (Hydrogen, its technology and implications;
    v. 4)
   Bibliography: p.
   Includes indexes.

1. Hydrogen. 2. Hydrogen as fuel. I. Cox,
Kenneth E. II. Williamson, Kenneth D.

III. Series.
TP359.H8H9 vol. 4 [(TP245.H9)] 665'.81'08s [(665'.81)]
ISBN 0-8493-5124-3   78-21146

This book represents information obtained from authentic and highly regarded sources. Reprinted material is quoted with permission, and sources are indicated. A wide variety of references are listed. Every reasonable effort has been made to give reliable data and information, but the author and the publisher cannot assume responsibility for the validity of all materials or for the consequences of their use.

All rights reserved. This book, or any parts thereof, may not be reproduced in any form without written consent from the publisher.

© 1979 by CRC Press, Inc.

International Standard Book Number 0-8493-5120-0 (Complete Set)
International Standard Book Number 0-8493-5124-3 (Volume IV)

Library of Congress Card Number 78-21146
Printed in the United States

PREFACE TO HYDROGEN: ITS TECHNOLOGY AND IMPLICATIONS

The United States, Western Europe, Japan, and several other countries are presently faced with an energy shortage due largely to an imbalance of energy consumption over fossil energy production. This problem was dramatized in October 1973 during the Arab embargo on the shipment of oil to the United States and the resultant large increases in the price of crude oil. This shortage in energy supply was then termed the "energy crisis." It was a clear demonstration of the nation's dependence on imported petroleum and its vulnerability on both political and economic grounds. It is clear that the above problems would worsen in the future unless more attention and effort are directed toward increasing domestic energy production from both depletable and non-depletable sources and reducing energy consumption.

In the short-term, until the year 2000, coal and nuclear energy are expected to play dominant roles in meeting the energy shortage despite the environmental restrictions that hamper the production and consumption of high-sulfur coal and similar difficulties (siting and radioactive waste disposal) that have slowed the devlopment of nuclear energy. In the long-term, beyond the year 2000, it is imperative that all forms of renewable energy be developed. These include solar energy, in such forms as wind, ocean thermal gradients, and biomass; geothermal energy; and fusion.

A major problem with several of the renewable energy sources is that they are intermittent and their energy density is low; thus, there is a need for an energy carrier that can act as both a storage and transportation medium to connect the energy source to the energy consumer. Many of the renewable energy forms, together with coal and fission exhibit their energy in the form of heat release. It is necessary to develop an energy carrier, other than electricity, to supply the transportation sector as well as overcome the problems of electrical storage.

Hydrogen, the lightest element, has been suggested as the energy carrier of the future. In itself, it is not a primary energy source but rather serves as a medium through which a primary energy source (such as nuclear or solar energy) can be stored, transmitted, and utilized to fulfill our energy needs. There are several distinct advantages to the use of hydrogen as an energy medium. It can be made from water, an inexhaustible resource. On combustion, water is the main product; thus, hydrogen can be regarded as a clean, nonpolluting fuel. Indications from current research efforts suggest that hydrogen may be produced from high-temperature heat sources at an efficiency greater than that of electrical generation, thereby making hydrogen a more economical energy source than electricity. Technology has already been developed for storing hydrogen as a pressurized gas, a cryogenic liquid, or in the form of a metal hydride. Systems for transporting hydrogen as a gas or a liquid have been developed with liquid hydrogen playing a major part in NASA's putting a man on the moon. Finally, hydrogen is of value as a chemical intermediate, being used in fertilizer manufacture, methanol synthesis, and petroleum treatment. This area of hydrogen utilization represents 3% of today's energy consumption and is expected to grow by a factor of five by the year 2000. The above concept of using hydrogen is termed the "hydrogen energy economy" and has been receiving an increasing amount of attention from energy scientists and engineers in the United States and abroad.

This series in five volumes represents a serious attempt at providing information on all aspects of hydrogen at the postgraduate and professional level. It discusses recent developments in the science and technology of hydrogen production: hydrogen transmission and storage; hydrogen utilization; and the social, legal, political, environmental, and economic implications of hydrogen's adoption as an energy medium. Although there are several reports of selected studies on hydrogen as a fuel, this is the first comprehensive reference book that covers a wide range of topics of notable interest and timely importance.

Volume I of the series discusses such topics as hydrogen production from fossil fuels, nuclear energy, and solar energy. Hydrogen production technology from water by traditional methods such as water electrolysis and newer attempts to split water thermochemically are included with details of current research efforts ad future directions.

Volume II provides detailed design information on systems necessary for storage, transfer, and transmission of gaseous and liquid hydrogen. Cost factors, technical aspects, and models of hydrogen pipeline systems are included together with a discussion of materials for hydrogen service. Metallic hydride gaseous storage systems for the utility and transportation industry are covered in detail, and the design Dewars and liquid hydrogen transfer systems are examined.

Volume III focuses on hydrogen's properties and provides in one location all of the hydrogen data measured and compiled by the National Bureau of Standards, Cryogenic Division. The properties are individually discussed, and tables of data are provided. The properties of slush hydrogen are also included.

Volume IV covers the present and future uses of hydrogen. Hydrogen has been suggested as a prime candidate for both air and surface transportation. In the utility industry, hydrogen systems for peak shaving promise to play an important future role. Both present and future domestic and industrial applications of hydrogen are surveyed. These include present uses in ammonia and methanol synthesis and future uses in the direct hydrogasification of coal to synthetic natural gas. Important to all of these applications are the safety considerations in the use of hydrogen to allow for public acceptance of hydrogen's role as an energy medium.

Volume V is primarily concerned with the nontechnical aspects of hydrogen. Economics of hydrogen energy systems will play a major part in determining the time frame for hydrogen's adoption. Cost analyses of such systems with return on investment considerations are surveyed from the point of view of production, transmission, and storage of hydrogen. The environmental, political, social, and legal implications of new secondary energy forms such as hydrogen are discussed with reference to governmental energy policy, the social costs of energy production and use, and the public's acceptance of a hydrogen energy medium.

The unusually broad nature of hydrogen demands the expertise that could only be provided by a wide authorship; thus, some of the authors are the original authorities in their respective fields. Although the subject matter treated in each chapter is, in general, the author's research work and his critical review of the state-of-the-art, the authors have had complete freedom in choosing the particular important areas to be emphasized. As a result, some chapters treat the subject matter in more detail than others with a greater emphasis on the engineering or design aspects of a particular system. Therefore, each chapter possesses its own special feature and appealing points. Due to the limited space in the series, the editors have encouraged each author to supply an extensive list of references at the end of his chapter for the benefit of interested readers. Detailed author and subject indexes have been provided at the end of each volume.

The editors, while striving to avoid duplication, have allowed some degree of overlap in certain of the chapters for the sake of continuity and allowing the reader to view a particular topic from two or more points of view. Further volumes on the topic of hydrogen are planned, and we wish to hear from our readers as to areas that might have been neglected or deserve a special chapter on their own.

We would like to express our sincere thanks to these authors and the staff of CRC Press, Inc. in particular Mrs. Gayle Tavens and Miss Sandy Pearlman, for their efforts in making these volumes possible. Lastly, we would like to thank our wives, Patricia R. Cox and Ruth S. Williamson, for their encouragement and help during the time it took to edit these five volumes.

K. E. Cox
K. D. Williamson, Jr.
Los Alamos, New Mexico
March 1975

PREFACE TO VOLUME IV: UTILIZATION OF HYDROGEN

This volume of *Hydrogen: Its Technology and Implications* deals with all facets of hydrogen utilization. In the first chapter, William VanVorst and Ronald Wooley discuss the application of hydrogen as a fuel for surface transportation. Following a brief history of hydrogen engines, the authors discuss engine cycles, special considerations for hydrogen fueled engines and vehicles, and prototype hydrogen vehicles using liquid hydrogen, fuel cells, and metal hydrides. The chapter is concluded with a brief discussion of vehicle safety and the application of hydrogen to railroad transport. In Chapter 2, Daniel Brewer presents the latest information on progress related to the development of both subsonic and supersonic hydrogen fueled aircraft. This chapter also includes a section on the history of hydrogen in aeronautics which covers everything from its use in airships to its use in the U.S. space program. Chapter 3 presents the current technical status of the distribution and residential use of hydrogen. In this chapter, J. Pangborn, M. Scott, P. Ketels, and J. Sharer discuss the compatibility of hydrogen to the existing natural gas residential distribution systems, residential use patterns, the conversion of existing domestic appliances, and the development of catalytic appliances. Daniel Cooperburg discusses the current and potential future industrial uses of hydrogen in Chapter 4 covering such processes as ammonia manufacture, oil shale liquifaction, and ethylene glycol production. The concluding chapter is an in depth treatise by F. J. Edeskuty on the subject of safety.

It has been a pleasure working with these experts in the field of hydrogen technology. The result is this up-to-date volume which we hope will encourage the application of hydrogen to the solution of some of society's energy problems.

K. D. Williamson, Jr.
Los Alamos
September 1978

THE EDITORS

Kenneth E. Cox, Ph.D., is a Staff Member in the High-Temperature Chemistry Group at the Los Alamos Scientific Laboratory, Los Alamos, New Mexico. He is currently performing research in developing practical thermochemical cycles that produce hydrogen by water-splitting. In his previous position as Professor of Chemical Engineering at the University of New Mexico, Albuquerque, he pioneered the development of photovoltaic-electrolytic methods to produce hydrogen from solar energy and water.

Dr. Cox graduated in 1956 from the Imperial College of Science and Technology of the University of London with B.Sc. and A.C.G.I. degrees in chemical engineering. He received his M.A.Sc. from the University of British Columbia in 1959 and his Ph.D. from Montana State University in 1962, also in chemical engineering.

Dr. Cox is a member of the American Institute of Chemical Engineers, the American Chemical Society, and Sigma Xi, the Scientific Research Society. He is also a member of the International Association for Hydrogen Energy and has contributed both as an author and reviewer to the *International Journal of Hydrogen Energy.*

Dr. Cox has published numerous research papers and has given over 100 presentations. His current research interests include: thermochemical generation of hydrogen from water, use of solar energy for hydrogen generation, techno-economic energy evaluations as well as traditional chemical engineering, e.g., thermodynamics of separation processes, process design, and dispersion phenomena in fluids.

K. D. Williamson Jr., Ph.D. is Assistant Division Leader of the Systems Analysis and Assessment Division, Los Alamos Scientific Laboratory, Los Alamos, New Mexico.

Dr. Williamson received his B.S., M.S., and Ph.D. degrees in chemical engineering from Pennsylvania State University in 1957, 1959, and 1961, respectively.

Dr. Williamson is a member and lecturer for the New Mexico Academy of Sciences and has served as an invited lecturer at the University of Tennessee and New Mexico Highlands University.

Dr. Williamson has published more than 35 research papers. His current interests include technology assessments and systems analyses of energy-related topics in support of research and development decision makers at Los Alamos and the State and National government levels.

CONTRIBUTORS

G. Daniel Brewer
Manager, Hydrogen Programs
Lockheed-California Company
Burbank, California

Dan Cooperberg
Process Planning Department
The Lummus Company
Bloomfield, New Jersey

F. J. Edeskuty
Associate Group Leader
Cryogenics Group
Los Alamos Scientific Laboratory
University of California
Los Alamos, New Mexico

Jon B. Pangborn
Director
Alternative Energy Systems Research
Institute of Gas Technology
Chicago, Illinois

Maurice I. Scott
Physicist-Project Engineer
Institute of Gas Technology
Chicago, Illinois

William D. Van Vorst
Professor of Engineering Systems
School of Engineering and Applied
 Sciences
University of California at Los Angeles
Los Angeles, California

Ronald L. Woolley
Technical Assistant to the President
Billings Energy Corporation
Assistant Professor of Mechanical
 Engineering
Brigham Young University
Provo, Utah

TABLE OF CONTENTS

Chapter 1

Hydrogen-fueled Surface Transportation

Chapter 1

HYDROGEN — FUELED SURFACE TRANSPORTATION

W. D. Van Vorst and R. L. Woolley

TABLE OF CONTENTS

1.1. INTRODUCTION

Several factors justify the current interest in hydrogen-fueled surface transportation. When combusted with air, hydrogen offers emissions free of the major contributors to atmospheric pollution: unburned hydrocarbons, carbon monoxide, carbon dioxide, oxides of sulfur, and greatly reduced amounts of oxides of nitrogen under normal operating conditions. In the "ultimate" system of combustion with pure oxygen, emissions would be truly pollution free. The use of hydrogen also promises reduced dependency on imported energy, as indigenous sources of energy would be used to generate hydrogen as an energy carrier and storage medium. All energy resources are potential feedstocks for hydrogen production. This includes renewable resources (solar, wind, falling water, tidal, ocean, thermal, biomass, etc.) and noncarbon-containing reserves (uranium, tritium, deuterium, and geothermal) as well as fossil reserves, most notably coal and oil shale. The magnitude of the problems of production and distribution of hydrogen must not be dismissed lightly and is dealt with in Volumes I and II of this series. Finally, experience with hydrogen-fueled engines to date indicates that conventional engines may be converted to hydrogen service without expensive retooling. Major problems with the use of hydrogen are those of preignition and flashback and suitable storage on board the vehicle.

1.2. HISTORY OF THE HYDROGEN ENGINE

The use of hydrogen as an engine fuel is not new, and, indeed, it fueled what may have been the first internal combustion engine. This was the engine developed in 1820 by the Reverend W. Cecil on the vacuum principle.[1] Hydrogen, burning with air, furnished a hot gaseous mixture which was expanded to atmospheric pressure and then cooled, resulting in a partial vacuum. The actual work was produced by the atmosphere forcing the piston back against the vacuum. The engine was said to have run satisfactorily, but vacuum engines in general never became practical and were supplanted by those in which power production was accomplished directly on the expansion stroke resulting from the increase in pressure due to combustion. As Cecil noted, however, "The engine, in which hydrogen gas is employed to produce moving force . . . may be inferior, in some respects, to many engines at present employed; yet, it will not be wholly useless, if, together with its own defects, it should be found to possess advantages also peculiar to itself."

It was a stationary engine, of course, and its "peculiar advantages", as claimed by Cecil, were that it could be located anywhere (in contrast to a waterwheel) and that it could be brought to full power very quickly (in contrast to the extensive warm-up period required for the steam engines of the day). More than 100 years later (1933), Erren and Hastings-Campbell[2] were to write in much the same vein, with the advantages claimed to be those of reducing atmospheric pollution and aleviating the dependence of England on petroleum imports. Now in the 1970s, we read essentially the same sentiment with the added advantage of increased thermal efficiency.

The literature also reveals an English patent application by Bursanti and Matteucci in 1854 for a free-piston type engine which would burn hydrogen. Lichty notes that a prototype was constructed by Benini in 1856.[3] Otto's engines of the 1860s and 1870s used substantial proportions of hydrogen in the gaseous fuels employed before the advent of gasoline. Interestingly enough (since safety is a natural concern in considering hydrogen as a fuel), Otto's first experiments with gasoline were so disastrous that he considered it much too unsafe and returned to the use of gaseous fuels.[4] Development of the carburetor permitted safe and practical usage of gasoline and, indeed, had the effect of almost terminating interest in any other fuel.

Interest was revived during World War I. Stimulating factors were threatened shortages of gasoline and potential military uses of hydrogen. These military uses included a hydrogen-oxygen system for submarines and use of hydrogen engines to propel airships. Much of this work was done in Germany by Erren whose notes and records were lost in the bombings of World War II. The work is not well documented, although it is alluded to by others, particularly Erren and Hastings-Campbell[2] and Weil.[5] Apparently, Erren converted a wide variety of engines to run on hydrogen and suggested direct injection as a solution to the preignition and flashback problems (to be discussed later). The submarine research was motivated by the possibility of a completely condensible exhaust from a hydrogen-oxygen engine. Weil also notes the interest during the 1920s of the possible use of hydrogen either as the primary fuel or as a "fuel-extender" for the airships then under construction in England and Germany for long-distance transportation of passengers and cargo. Since hydrogen provided the necessary buoyant force and must be partially vented during the voyage, there would be no added storage problem. It was an appealing and promising suggestion.

The first rigorous, experimental investigation of the performance of an internal combustion engine fueled with hydrogen appears to have been undertaken by Ricardo[6] in England. Using single-cylinder, variable compression engines, he investigated combustion over a wide range of hydrogen-to-air mixture strengths. He reported high thermal efficiencies generally, obtaining 43% at a compression ratio of 7 at an unspecified, but lean, mixture strength. He also reported the problems of preignition and flashback, or "popping-back into the carburetor", which prevented operation at mean effective pressures greater than 74 psig (510 kPa).

Also in England, Burstall[7] in 1925 reported his observation of these problems. He reported an indicated thermal efficiency of 41.3% at a compression ratio of 9.95 and an equivalence ratio (the actual hydrogen-air ratio used compared to the stoichiometric) of 0.587. He was unable to exceed an equivalence ratio of 0.80.

During World War II, Australia became seriously short of petroleum-based fuels as a result of the wartime blockade. This stimulated intensive effort on the development of alternative fuels, including hydrogen. Prototype vehicles were successfully operated[8] before the end of the war, although with the consequent decrease in the need for "other" fuels, no further development was pursued.

Work on hydrogen-fueled engines also continued in Germany through World War II. In reviewing hydrogen engine work generally, de Boer et al.[9] have called particular attention to the work of Oemichen[10] who investigated the use of direct cylinder injection of hydrogen at length.

After the war, King in Canada conducted an extensive study of engine fuels in general and hydrogen in particular. His interest in the latter had been stimulated by the report of Erren and Hastings-Campbell;[2] he was present at the meeting in which their paper was presented and participated quite actively (and somewhat antagonistically) in the discussion. His investigations were directed specifically toward the flashback and preignition problems, and he advanced the general cause-and-effect relations that continue to be generally accepted. King and Rand showed[11] that preignition was caused by "hot spots" in the combustion chamber and by particulate matter such as would result from the pyrolysis of lubricating oil vapor or even "inert" dust in the intake air. Combustion knock was shown to be an inherent property of near stoichiometric hydrogen-air mixtures due to the high flame velocity.[12,13] King was able to completely eliminate preignition and flashback in the intake system by excluding fluffy carbon deposits and by using a "cold" spark plug and an "aged" sodium-filled exhaust valve. He achieved 47% indicated thermal efficiency at an equivalence ratio of 0.40 and a compression ratio of 12:1 without flashback as long as excessive oil was excluded from the combustion chamber.[11]

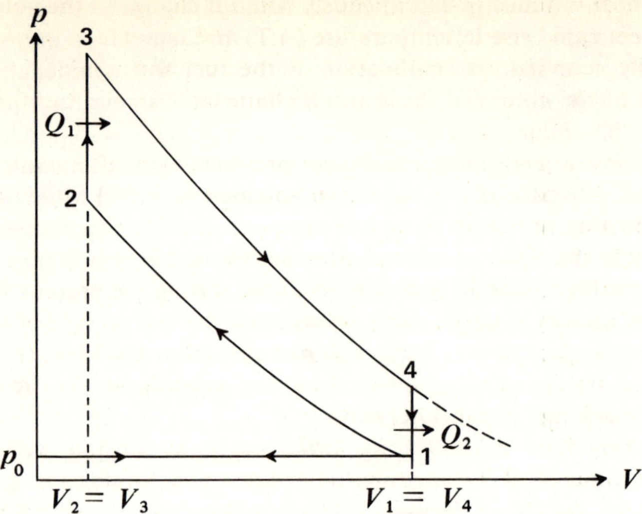

FIGURE 1.   The idealized Otto cycle.

It is also notable that a jet engine operating on hydrogen was flight tested in the 1950s,[14] and highly successful operation reported.

The foregoing may be considered as "past history" at this point. In the main, it has pointed up the severity of the problems of flashback and preignition with the combustion of hydrogen in an internal combustion engine. It is appropriate now to explore the nature of hydrogen as a fuel, the problems associated with using it, and the question of safety.

1.3. ENGINE CYCLES

A combined review of the most common engine cycles and the general combustion properties of fuels leads to an appreciation of the advantages and disadvantages of hydrogen as a fuel for internal combustion engines.

1.3.1. The Air Standard Otto Cycle

By far the most common of the numerous engine cycles currently in use is that generally attributed to Otto (1876). It forms the basis of the ordinary gasoline engine and is shown schematically in Figure 1. It is advantageous to consider first that only air undergoes the series of processes implied, then discuss the effects of the fuel.

1.   The charge is slowly drawn into the cylinder, ideally at constant (atmospheric) pressure (Process 0-1). In practice, of course, this is done rapidly, and the moving piston creates a partial vacuum in the cylinder. This departure from ideality, however, is entirely a matter of mechanical design and has no influence on the choice of fuel.

2.   With the intake valve closed, the piston compresses the gas adiabatically from V_1 to V_2, with a consequent rise in temperature and pressure. The compression ratio, V_1/V_2, is a fundamental engine design variable and, as will be shown later, is limited by the nature of the fuel (as well as by the ultimate structural considerations). The departure from the ideal here is largely a result of intake valve timing, ring blow-by, and heat transfer — again, variables not influenced by the nature of the fuel.

3. Ideally, heat is added instantaneously without change in the volume V_2, with the consequent rapid rise in temperature to T_3 and increase in pressure P_3. The heat is actually supplied via combustion of the fuel and nonideality is very much a function of the nature of the ignition characteristics and burning velocity of the fuel.

4. The gas, now at high temperature and pressure, expands against the piston, driving it back adiabatically to the initial volume ($V_1 = V_4$) while undergoing significant reduction in temperature and pressure. Ideally, this expansion ratio should be precisely the same as the compression ratio, but it is not in practice because of valve timing. Since all power is produced during the process $3 \rightarrow 4$, this departure from ideality is particularly severe, and the critical role of the fuel is apparent. There are, certainly, further departures from the ideal resulting from heat losses and frictional effects, but these are subordinate to the effect of the reduced, ''working'' expansion ratio.

5. The working fluid is cooled, instantaneously, at constant volume. Practically, this is accomplished by opening the exhaust and forcing the gases out into the atmosphere. As far as the heat effect is concerned, this overall process is thermodynamically equivalent to that of constant volume cooling. Mechanically, however, the exhaust valve must open before point 4 is reached, thus further reducing the work possible during the expansion stroke; further, the exhaust stroke must begin at a pressure somewhat greater than atmospheric, thus detracting further from the net work deliverable by the cycle. Again, these nonideal effects are inherent design problems indendent of the choice of fuel.

Ideally, the effects of the intake and exhaust stroke are equal and opposite and cancel each other, leaving the net cycle represented by the simple ''four-cornered'' diagram 1-2-3-4-1. While this cannot be accomplished in practice and the nonidealities distort the closed diagram to resemble something like a malformed egg, analysis of the four-cornered diagram is useful in exploring certain aspects of fuel selection. The ''air-standard analysis'' of elementary engineering thermodynamics,[15] based on the idealizations set forth above and treating the working fluid as having the properties of air throughout the cycle, leads to the following relation for the efficiency of the cycle:

$$\eta = \frac{1}{r^{\gamma - 1}} \qquad (1)$$

where r is the compression ratio, V_1/V_2, γ is the ratio of specific heats, and η is the thermal efficiency.

1.3.2. Comparison of Fuel Properties

Table 1 presents the basic combustion properties of several fuels and serves as a basis for discussing the influence of the fuel on the efficiency of the cycle. More extensive tables of hydrogen properties are contained elsewhere in this set of volumes.

While properties of several fuels are shown, it is the comparison of hydrogen and gasoline that is of most immediate interest. It will first be noted that the autoignition temperature of hydrogen is the highest of all the fuels shown and is significantly higher than that of gasoline. This has important implications for the ordinary engine in which a combined charge of fuel and air is compressed. Again utlizing the methods of elementary thermodynamics,[15] the temperature rise during compression is found to be exponentially dependent upon the compression ratio:

TABLE 1

Combustion Properties of Several Fuels (Generally Accepted Values from the Literature)

	H_2	CH_4	C_3H_8	Gasoline	CH_3OH
Autoignition temperature (°C)	585	540	510	440	385
Minimum ignition energy (MJ)	0.02	0.28	0.25	0.25	—
Flammability limit (vol % in air)	4—75	5—15	2.2—9.5	1.3—7.1	6.7—13.6
Stoichiometric mixture (vol % in air)	29.6	9.5	4	1.7	12.3
Maximum flame velocity laminar (cm/sec)	270	38	40	30	—
Diffusivity in air (cm²/sec)	0.63	0.2	—	0.08	—
Lower heat of combustion					
1000 Btu/lb	51.6	21.5	20	20	9.1
1000 Btu/gal	31.0[a]	76.1[a]	96.6[a]	120	60.0

[a] As saturated liquid at atmospheric pressure.

$$T_2 = T_1 \left(\frac{V_1}{V_2}\right)^{\gamma - 1} \tag{2}$$

The compression ratio is limited then by the necessity of insuring that T_2 does not increase to the autoignition temperature thus permitting premature combustion and rough, unsteady operation. While the ratio of specific heats, γ, is influenced by the fuel, the effect is small due to the comparatively small fraction of fuel present. The temperature of the charge at the onset of compression is not unique, of course, depending on the atmospheric condition and heat transfer characteristics of the engine. Moreover, the autoignition temperature is not well defined numerically so that, all in all, estimates of the permissible compression ratio from Equation 2 (with T_2 set equal to the autoignition temperature as a maximum) are exploratory at best. Still, doing so gives figures of interest for purposes of comparison. If a given set of conditions indicates a limiting compression ratio of 9 for an engine using gasoline as a fuel, for example, 14 might be indicated for the engine with hydrogen as a fuel. From Equation 1 then, the thermal efficiency would be expected to increase accordingly.

The second line of Table 1 shows the order of magnitude lower minimum ignition energy of hydrogen as compared to other fuels. While this is a generally desirable attribute, it also contributes to the problems of preignition and flashback noted by Ricardo, Burstall, and others, as noted earlier. Apparently, so little energy is required that any residual hot spot in the intake system or cylinder or hot particulate matter may serve as a source of ignition, thus leading to these phenomena.

Lines three and four of Table 1 should be examined together, as they indicate perhaps the most critical property — that of the mixture ratios necessary for combustion. If combustion takes place at or very close to the stoichiometric condition, the highest temperature possible (due to combustion) is reached. This influences the composition of the products of combustion, as well as imposing the severest demands on the materials used in the fabrication of the engine. In particular, it enhances the formation of oxides of nitrogen, of CO at the expense of CO_2, and unburned hydrocarbons (UHC). In general, it is not the preferred combustion ratio, and a mixture on the ''lean'' side is more desirable. This is the condition of less than the stoichiometric amount of fuel for a given amount of air, i.e., oxygen rich. This mode generally results in more complete combustion and greater fuel economy. Note, however, that with gasoline the flammability limits imply that combustion can take place at a fuel composition only slightly less than the stoichiometric, 1.3 compared to 1.7% fuel by volume. With hydrogen, on the other hand, while the stoichiometric ratio calls for 29.6% fuel in the air, combustion may occur with as little as 4%. This permits relatively low temperature combustion with minimal formation of oxides of nitrogen. Since there is no concern with carbonaceous products, pollutants are virtually eliminated from the products of combustion. This too, however, like the minimum ignition energy figures, must be recognized as a mixed attribute, for the wide range of flammability limits also allows undesired combustion, contributing to the problems of preignition and flashback.

The flame velocity is substantially greater for hydrogen than for other fuels and is almost an order of magnitude greater than that for gasoline. This should allow combustion with hydrogen to approach the ideal, constant volume process of Figure 1, (2-3) much closer than would be possible with gasoline. This potential advantage must be viewed with considerable caution, however, as the figures shown are for mixture ratios near the stoichiometric for each fuel. As de Boer et al. have pointed out,[9] the flame velocity decreases significantly as the mixture ratio decreases from the stoichiometric. It seems likely that the velocities might be comparable for the lean combustion of hydrogen and the near-stoichiometric combustion of gasoline.

The "diffusivity" is also shown for several fuels in air. The significantly higher value for hydrogen indicates two potential advantages. First, mixtures of hydrogen in air should reach a higher degree of uniformity of composition in the time available than those of gasoline in air. Second, and more of a safety advantage, any leaking hydrogen would disperse rapidly and would be carried away into the atmosphere, while gasoline tends to remain and concentrate, thus constituting a severe explosion hazard.

The last two lines of Table 1 indicate heats of combustion of the several fuels. It will be noted that hydrogen has more than twice the value of gasoline on a weight basis; however, the density of hydrogen is so low that the picture is more than reversed on a volume basis. Since most vehicles are inherently "volume limited", storage space per se is a serious problem with hydrogen, even in the liquid state.

1.3.3. The Air Standard Diesel Cycle

Most of the above discussion is directly applicable as well to the diesel cycle, shown in Figure 2. The diesel cycle differs from the Otto most significantly in that combustion takes place, ideally, at constant pressure subsequent to compression rather than at constant volume. This is accomplished by continuous injection of the fuel over a short time interval rather than mixing the charge prior to compression. Thus, the expansion ratio, V_4/V_3, differs from the compression ratio, V_1/V_2; although the cylinder volume is, in fact, expanding during combustion, this is usually considered just the combustion process, since the *adiabatic* expansion cannot begin until combustion is finished. Non-idealities of the actual engine processes are much the same as for the Otto. The actual combustion process is closer to the ideal in the diesel, and autoignition is avoided until desired by injecting the fuel.

Analysis of the idealized, four-cornered diagram of the diesel cycle[15] yields

$$\eta = 1 - \frac{1}{r^{\gamma - 1}} \left[\frac{(r_c - 1)}{\gamma (r_c - 1)} \right] \tag{3}$$

where r_c is the so-called cutoff ratio, i.e., the ratio of V_3 to V_2, V_3 being the cylinder volume at the instant fuel injection occurs.

Comparison of Equations 3 and 1 shows that, for the same compression ratio, the Otto cycle is more efficient than the diesel, provided the cut-off ratio of the diesel, r_c',

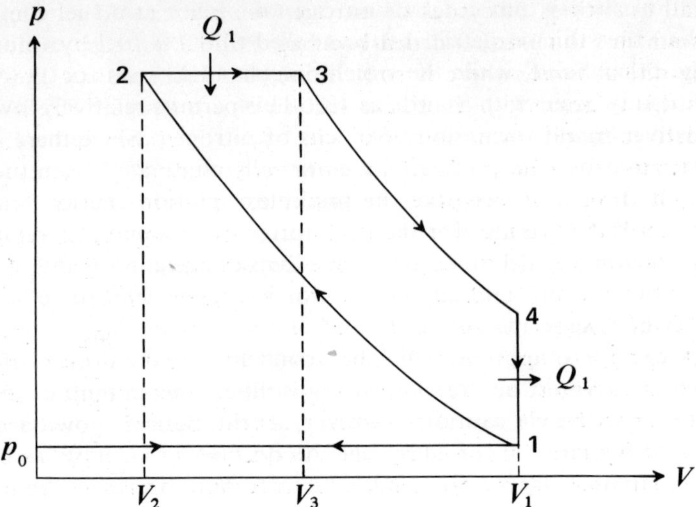

FIGURE 2.   The idealized diesel cycle.

is greater than 1 — as it invariably is in practice. The reputation of higher efficiency of the diesel cycle then comes about through its operation at higher compression ratios than are practical with the Otto cycle. This characteristic need for high compression ratios of the diesel engine implies certain problems associated with hydrogen-fueled operation. With the ordinary hydrocarbon fuels, smooth ignition results from the injection of the fuel into the comparatively high-temperature, high-pressure air after compression. With hydrogen, this type of compression ignition apparently does not occur. de Boer et al.,[16] using a diesel injection head on a CFR engine, report no such ignition with hydrogen at compression ratios as high as 29, where the absolute temperature rise ratio of the air during compression should approximate 3.5, i.e., for an initial air temperature of 70°F (21°C), the temperature after compression should be 1395°F (758°C). Thus, some sort of ignition source must be used. de Boer reports that ignition is readily accomplished by means of a simple "glow-plug" system. It should also be possible to inject small amounts of hydrocarbon fuel which would autoignite and serve as an ignition source for the hydrogen. The problem does not seem a severe one, therefore; a more substantive problem is that of injection of gaseous hydrogen. In the case of liquid fuels, there is little difficulty in obtaining the high pressure required and introducing the required amount of fuel into the cylinder within the desired time frame. With hydrogen, however, compression of the gas to the high pressure required would involve a severe energy penalty and the development of a suitable compressor. Storage at high pressure does not seem an acceptable alternative, from weight as well as general safety considerations. Messerole and de Boer[17] have proposed a pump designed to accomplish pressurization in the liquid phase, which would reduce the energy required to an acceptable level. At the time of the injection proper, however, the hydrogen would be gaseous, and the extremely low density imposes severe design constraints on the injector system — that is, a comparatively large volume of gas must enter the cylinder in a very short time.

While the problems of ignition and injection are not to be dismissed lightly, they do appear solvable by intensive developmental effort. Granting such, hydrogen offers substantial advantages as a fuel for diesel engines. With hydrocarbon fuels, diesel engines are almost prohibitively productive of atmospheric pollutants in the exhaust products. These would be virtually eliminated with hydrogen, of course, which would yield only small quantities of oxides of nitrogen when operated very lean or slightly rich; trace amounts of the usual hydrocarbon base pollutants might be generated from the lubricating oil, although there is some hope that lubricants of the future might eliminate even this minor effect. In a later section of this chapter, it is shown that the engine could deliver more power per stroke with hydrogen. The combustion properties of hydrogen discussed earlier indicate an improved thermal efficiency of the diesel cycle fueled by hydrogen as compared to gasoline, and some corraborative evidence has been reported.[16,18] Operation at lean mixtures was reported as very smooth, although preignition roughness was noted at, or near, stoichiometric conditions.

1.3.4. Other Engine Cycles

Hydrogen-fueled operation should also be considered for the Joule or Brayton cycle. This cycle finds its greatest application in the modern gas turbine or jet propulsion engine. The cycle may be visualized as similar to the diesel, with the constant-volume cooling process of the latter replaced by one of constant pressure. Flow is continuous in such engines, eliminating the problems of intake- and exhaust-valve operation. As de Boer et al.[9] note, such engines appear to be particularly suitable for hydrogen fueled operation, and, as noted earlier, successful flight tests of jet propulsion engines have been reported.[14] Again, greatly reduced pollution and improved efficiency obtainable through lean operation with hydrogen may be expected.

de Boer et al.[9] give references to extensive work at the NASA-Lewis Research Center; while emphasis is, of course, on aircraft applications, the possible use of gas turbines for surface transportation should not be ignored. Such usage on a large scale is inhibited, however, by the greater expense of such engines (in comparison with either diesel or Otto engines) and by the limit of the maximum continuous temperature permitted by the materials available (on a practical basis) for the turbine nozzles and blades. This, in turn, limits the efficiency, which ideally should be identical with that of the Otto cycle for the same compression ratio.[15] Operating efficiency of the gas turbine may be enhanced considerably by the use of recuperative heat exchange, however, and the successful development of this process would make the gas turbine competitive for automotive usage.

Since the fuel is injected directly into the combustion chamber of the gas turbine engine, the problems of adapting to hydrogen fueled operation are essentially those discussed above for the diesel. They should be somewhat simpler to deal with, however, as glow plugs are commonly used with gas turbines and are simple to incorporate in their design. The injection problem should be far simpler, since fuel admission would be continuous; thus, pressurization is the only problem — there being no need for the complex timing required for injection in a reciprocating engine. The lower pressure ratios of the gas turbine also inherently simplify the injection process.

The intent of this section has been to discuss the most frequently used engine cycles and to show rather generally that their performance would be enhanced with hydrogen as the fuel. It is appropriate at this point to discuss the special problems of hydrogen powered engines that have been encountered by researchers.

1.4. SPECIAL CONSIDERATIONS FOR THE HYDROGEN ENGINE

With the exception of rocket motors used in the space program, no engine has been designed specifically for hydrogen fuel at the present time. Special conditions therefore need to be taken into account when an engine that was designed for another fuel, such as gasoline, is converted for hydrogen operation. The essential difference in the conversion is that the engine is now a ''dry fuel'' type in the general category with propane and natural gas engines. Conversion techniques and precautions that apply to propane and natural gas engines also apply, in general, to the hydrogen engine. Therefore, the literature on these other gaseous fuels provides a useful source of information on hydrogen engines as well. Both propane and natural gas regulators and carburetors have been used in many conversions, usually with only minor modification, such as change of spring, to accommodate the lighter gas. The modifications to existing equipment have been minor because of a fortunate circumstance: the flow-reducing low density of hydrogen is somewhat compensated for by a lower frictional resistance to flow and higher speed of sound (choked flow condition occurs at much higher velocity).

Ignition timing is generally simplified with hydrogen. Designs that use a fixed equivalence ratio can use fixed timing at all engine revolutions per minute and manifold vacuum settings. Spark timing is nearly independent of these two parameters. In a variable equivalence ratio system, proper timing is a function of the equivalence ratio with the greatest advance required for lean operation (Figure 9).

Engines operating on hydrogen have a greater tendency for the fuel-air mixture to ignite in the intake manifold than do engines with other gaseous fuels. This topic is covered in detail below. The power and efficiency of hydrogen fuel differs substantially from other fuels and is strongly dependent on engine design.

1.4.1. Preignition and Flashback

The low energy of ignition and the wide flammability limits of hydrogen have been noted earlier as attributes of hydrogen as an internal combustion engine fuel. Coincidentally, however, these properties also lead to the problems of preignition and flashback historically associated with such use of hydrogen. Preignition, as the name implies, notes ignition of the fuel-air mixture prior to the time at which it is desired. "Flashback" refers to an extreme condition, in which ignition may take place in the intake system or before the intake valve closes, so that a "firing back" through the intake system takes place. This may result in damage to the device in which fuel and air mixing takes place and may interrupt or stop engine operation. Ricardo[6] and Burstall[7] clearly recognized the problem, and King and his co-workers[11-13] made a major experimental study of it. They reported being able to eliminate the problem by maintaining all surfaces (cylinder wall, piston face, etc.) scrupulously clean. This seemed to support their explanation that the phenomenon is caused by a mixture of hydrogen and air of ignitable composition coming into contact with some ignition source, such as a hot spot, or some particulate matter; they had concluded earlier that hydrogen does not autoignite, so that compression ignition could be ruled out. Other experimenters have confirmed that particulate matter, such as oil blow-by, contribute to flashback.

Billings et al.[19] have suggested that preignition may be initiated by a premature spark discharge due to a "voltage crossover" or induction from one spark plug lead wire to another. Shielded cables, nonparallel wire runs, improved insulation, or modification of the distributor rotor so that all plugs (except that being fired) are grounded would seem to eliminate this source. The spark plug electrodes, per se, may also serve as a source if they retain sufficient energy after firing to maintain an elevated temperature. This is apparently alleviated to a considerable degree by narrowing the gap to one third or one fourth of the normal distance. Gap reduction is also used to reduce the spark energy requirement at high compression. The effect is pronounced, with a much smoother operation resulting;[20] this would seem consistent with the lower energy required for ignition with hydrogen.

Two techniques have proved effective in combating or suppressing preignition and flashback in engines in which a combined fuel-air charge is admitted to the cylinder.

Finegold et al.[21] employed exhaust gas recirculation (EGR), recycling as much as 35% of the exhaust, after substantial cooling. This serves as a "thermal diluent", reducing peak combustion temperatures (and consequently, NO_x formation) and, apparently, lowering the temperature of potential hot spot ignition sources. The process is demonstrably effective, although it must be noted that power reduction is roughly in proportion to the percentage EGR employed.

The injection or induction of water may be employed to accomplish the same objective, as reported by Billings and Lynch[22] and investigated in more detail by Woolley.[23] Again, thermal quench of exhaust gas particulates and a general evening out of temperatures appear to account for preignition control and greatly reduced NO_x. The control system is far more flexible with this procedure, and water may be inducted only when needed. This can be a relatively small fraction of the time, as flashback control becomes important primarily at higher power demand when the equivalence ratio setting is high and the throttle if (any) is near the wide-open position. As used here, the equivalence ratio is defined as the actual fuel-air ratio employed to that called for by the balanced chemical reaction equation for the combustion process, i.e., the stoichiometric. Thus, at normal operation, the equivalence ratio might be 0.5:0.65, and little tendency to flashback is noted. At higher equivalence ratio settings (0.8:1.0), flashback control has been essential in all studies to date. With proper control, then, water may be inducted only as the equivalence ratio approaches unity, greatly reducing the water that need be carried on board the vehicle for this purpose. It may, indeed, prove feasible to condense sufficient water vapor from the exhaust to provide the liquid water required for induction.

While the "suppressive" measures of EGR and water induction successfully reduce the problem of preignition and flashback to a tolerable level, it seems doubtful they can eliminate it completely. The wide flammability limits and low ignition energy requirement of hydrogen-air mixtures seem to assure the possibility of preignition under a wide variety of operating conditions. Far more basic research on the nature of the ignition of hydrogen is required to identify the problem rigorously and, thus, to suggest the ultimate solution. Clearly, however, the problem would be avoided if the hydrogen were prevented from mixing with air until the instant ignition is desired. Thus, if hydrogen were injected after the air intake valve were closed, there could be no "firing back" through the intake system; if it could be injected just as ignition were desired, there would be no occasion for preignition. Such instantaneous injection is impossible, of course, and the problem of injection itself, is a difficult one, as implied under the discussion of the diesel engine, above. Nevertheless, fuel injection offers the possibility of true elimination of the preignition problem, and several techniques have been investigated.

Murray et al.[24] have developed hydrogen injection techniques in small-displacement, one-cylinder industrial engines. The operating cycle has been modified from a conventional Otto cycle to a throttled four-cycle spark ignition, constant pressure combustion cycle. By the strategic injection of the hydrogen, the possibility of preignition is circumvented. However, a comparatively high fuel injection pressure — up to 66 atm at the highest power output — is required. Reported performance was good, and the NO_x emissions were a small fraction of those obtained with a gasoline.

Adt and co-workers[25,26] have operated an automobile on hydrogen, using a specially designed intake valve. In essence, unthrottled air is inducted along with hydrogen from a special port installed in the intake valve seat. The back face of the head of the intake valve covers the hydrogen supply tube when the valve is closed and permits hydrogen injection when it is open. This minimizes "pumping losses" of the ordinary throttled intake system. While the pressure demanded is not great, there appear to be problems of flow and mixture ratio control.

MacCarley and Van Vorst[27] have attempted to inject the hydrogen just after the intake valve closes, while the cylinder pressure remains low; thus, high injection pressures would not be required. Great difficulty was encountered in devising an injector that could deliver the required amount of hydrogen in the time available at low injection pressures, near 5 atm. On smaller scale tests of a motorcycle engine, flashback was indeed eliminated, although preignition continued. The effect was minor, however, being reflected as roughness in operation.

Homan and co-workers[28] injected hydrogen directly into the cylinder of an ASTM-CFR engine near top dead center using a glow plug for ignition. They reported significantly higher indicated mean effective pressures (IMEP). IMEP on the order of 1.3ϕ MPa were observed for hydrogen with equivalence ratios ϕ between 0.1 and 0.4 at a compression ratio of 18. The corresponding result for diesel oil was about 0.6ϕ MPa for $0.3 < \phi < 0.9$. Increased thermal efficiency was also observed. Hydrogen indicated thermal efficiency was approximately 40% over the equivalence ratio range. This compared to 20 to 25% for diesel oil in the same test engine. NO_x formation was reported as 50 to 100 ppm or 0.2 to 0.4 g/MJ, NO_x emission was thus considerably higher than with premixed hydrogen-air design for the same lean combustion. Rich combustion in the region near the injection port was thought to be the cause.

It must be acknowledged, then, that the problem of preignition and flashback is a serious one. Apparently, it can be suppressed satisfactorily by EGR or water induction (probably the more effective) and, in principle, solved by injection of the hydrogen directly into the cylinder. No vehicle operation employing the latter has been reported to date.

1.4.2. Power and Efficiency

It is well known that a gasoline engine of a given displacement will have less power capability when running on hydrogen because the volume flow of gaseous hydrogen reduces the flow of intake air. The result is a reduction in the effective displacement of the engine. This drop in peak power output exists for all designs in which hydrogen mixes with air prior to intake valve closure. Examples are premixed designs using propane or natural gas carburetion equipment and direct cylinder injection designs where the hydrogen is injected prior to intake valve closure. The power of the hydrogen engine may be estimated from equations predicting the fuel flow rate at a given revolutions per minute.

$$v_{H_2} = \left[\frac{\dot{m}_{air}}{\rho_{0,\,air}\; x_{air}\; N\; V_d/2} \right]_{H_2} \tag{4}$$

$$\dot{m}_{H_2} = \rho_{H_2}\; Q_{H_2} = \rho_{H_2}\; \frac{\dot{m}_{air}}{\rho_{air}}\; \frac{x_{H_2}}{x_{air}} \tag{5}$$

$$x_{H_2} = \frac{2\phi}{2\phi + 4.762} \tag{6a}$$

$$x_{air} = \frac{4.762}{2\phi + 4.762} \tag{6b}$$

$$\rho = \frac{p\,w}{R_u\,T} \tag{7}$$

$$\dot{m}_{H_2} = \left(\left(\frac{N\,V}{2}\,d \right) \left(\frac{2\phi}{2\phi + 4.762} \right) \rho_0\; \frac{w_{H_2}}{w_{air}}\; v \right)_{H_2} \tag{8}$$

$$P_{H_2} = \dot{m}_{H_2} \, \Delta h_{H_2} \, e_{H_2} \tag{9}$$

A definition of volumetric efficiency for a hydrogen engine may be constructed analogous to the conventional definition which is the ratio of air mass flow rate divided by the swept volume times the density of air at the reference pressure and temperature (P_o, T_o). The volumetric efficiency characterizes the frictional and pumping characteristics of the intake system as well as flow reduction due to heating of the air along the intake flow path. Therefore, an analogous definition is the actual mass flow rate of entering air divided by the mass of air at P_o, T_o that would fill the swept *partial* volume of the engine (Equation 4).

The mass flow rate of hydrogen may be related to the air flow rate through the mole fraction (Equations 5 and 6). These equations are combined with the perfect gas law (Equation 7) to provide an estimate of hydrogen usage (Equation 8) at wide open throttle and the power capability of the engine (Equation 9).

Engine revolutions per minute may be related to road speed (Equation 10).

$$N = \frac{G_R \, V}{\pi \, T_d} \tag{10}$$

The overall gear ratio enters the power formula as a design parameter that can be used to compensate for reduced power capability by running the engine faster at a given road speed. In this way, acceleration performance can be regained at a sacrifice of top speed. An added benefit of increased gear ratio is a lessening of the tendency to flashback under load.

The power output of a gasoline engine can be estimated in a similar manner by the equations given above. Equation 11 is the standard definition of volumetric efficiency.

$$v_{gaso} = \left(\frac{\dot{m}_{air}}{\rho_{0,\,air} \, N \, V_d/2} \right)_{gaso} \tag{11}$$

$$\dot{m}_{gaso} = \left[\frac{\dot{m}_{air}}{A_f} \right]_{gaso} \tag{12}$$

$$P_{gaso} = \dot{m}_{gaso} \, \Delta h_{gaso} \, e_{gaso} \tag{13}$$

The equations for engine brakepower capability at a given road speed may be ratioed to show the power reduction of a converted engine and the terms that affect the power (Equation 14).

$$\frac{P_{H_2}}{P_{gaso}} = \left(\frac{\Delta h_{H_2}}{\Delta h_{gaso}} \right) \left(\frac{w_{H_2}}{w_{air}} \right) \left(\frac{v_{H_2}}{v_{gaso}} \right) \left(\frac{2\phi \, A_f}{2\phi + 4.762} \right) \left(\frac{e_{H_2}}{e_{gaso}} \right) \left(\frac{G_{R,H_2}}{G_{R,gaso}} \right) \tag{14}$$

Using a typical lower heating value for gasoline, Equation 14 reduces to Equation 15.

$$\frac{P_{H_2}}{P_{gaso}} = 0.191 \left(\frac{v_{H_2}}{v_{gaso}} \right) \left(\frac{2\phi \, A_f}{2\phi + 4.762} \right) \left(\frac{e_{H_2}}{e_{gaso}} \right) \left(\frac{G_{R,H_2}}{G_{R,gaso}} \right) \tag{15}$$

Power is seen to be altered by the volumetric efficiency, lean combustion, engine efficiency, and gear ratio. For a typical setting of $\phi = 0.7$ and $A_f = 15.1$, the converted engine will have 66% of the power of the gasoline counterpart if the volumetric efficiency, engine efficiency, and gear ratio remain the same. Thus, the challenge to the designer is to improve volumetric efficiency and engine efficiency as much as possible and then compensate with gear ratio change without overspeeding the engine.

Volumetric efficiency may be increased by using outside air in place of engine compartment air and by reducing air preheating when possible. Engine efficiency may be increased by compression ratio change and proper spark timing.

The above estimate is applicable to an engine design that combines hydrogen and air before closure of the intake valve. In an engine supplied with hydrogen at sufficient pressure for injection directly into the cylinder after the intake valve has closed, it is possible to have greater power output than with gasoline. In this case, the same equations apply with the exception that the mass of air entering the cylinder of the hydrogen engine is given by Equation 11 rather than Equation 4. Hydrogen no longer reduces the flow of air into the engine. It is now possible to have greater power output, since the energy release of the hydrogen-air reaction is greater than the gasoline-air reaction for the same amount of consumed oxygen.

The equation for power ratio of a hydrogen engine with direct cylinder injection to a gasoline engine is given as Equation 16.

$$\left(\frac{P_{H_2}}{P_{gaso}}\right)_{DCI} = 0.191 \left(\frac{v_{H_2}}{v_{gaso}}\right)\left(\frac{2\phi A_f}{4.762}\right)\left(\frac{e_{H_2}}{e_{gaso}}\right)\left(\frac{G_{R,H_2}}{G_{R,gaso}}\right) \qquad (16)$$

It is only slightly different from Equation 15 for a premix hydrogen engine design. If the converted engine has the same volumetric efficiency and engine efficiency as the gasoline engine and if no change in overall gears ratio is made, then the hydrogen engine is capable of 21% greater power output (A + = 75.1, ϕ = 1.0).

There are three problems encountered with direct cylinder injection of hydrogen: (1) mechanical complexity of the injector, (2) supply of pressurized hydrogen, and (3) increased NO_x. Much of the work in this area is being conducted by Daimler Benz[29] and at Cornell University.[28] The University of Miami and Hawthorn Research[30] are planning to study direct cylinder injection in the second phase of a 3-year study that began in 1977. The first phase considered premix designs that operate with a throttle valve in the manner of a conventional propane system. They have observed brake thermal efficiencies of around 30 to 33% over a wide range of equivalence ratios. Intake volumetric efficiency is quite similar to or slightly greater than that of a corresponding gasoline engine. The actual efficiency of a given engine depends on many design parameters, perhaps the most important of which is the compression ratio. Judging from the comparison of the data from several research groups as found in the first year report of the ERDA sponsored study,[30] it would seem that a reasonable estimate of brake thermal efficiency is 30% for a compression ratio of 9:1.

An analysis of the several strategies for design of the intake system of the hydrogen engine was made by Donnelly et. al.[31] as part of a theoretical comparison of hydrogen and battery vehicles. The results of the engine cycle analysis are given in Table 2. The comparison with battery vehicles is given in a separate section.

Daimler Benz has conducted experimental tests of premix and direct cylinder injection designs of a hydrogen engine.[29] Table 3 shows the theoretical effects of various mixture formation methods on performance and efficiency. Methods of controlling flashback using exhaust gas recirculation and water induction are compared. These methods are necessary for successful operation of premix or external mixture formation systems. Two types of direct cylinder injection techniques were estimated in Table 3 and then tried experimentally: late injection after intake valve closure and early injection prior to intake valve closure.

Tests were conducted on two engines: a single-cylinder engine with combustion chamber in the piston crown (displacement 0.614 ℓ, bore 92 mm, stroke 92.4 mm) and

TABLE 2

Results of Hydrogen Engine Cycle Analyses

	Naturally aspirated ICE	Direct cylinder injection ICE	
		Compressed hydrogen	Pumped hydrogen
Inlet hydrogen pressure and temperature (psia/°F)	14/80	14/80 14/200	14/≥ −423
Temperature of inlet mix (°F)	108	— —	—
Ideal cylinder charge density (lb/ft³)	0.0482 (mixture)	0.0652 (air) 0.0652 (air)	0.0652 (air)
		0.0671 (corr. for H_2) 0.0671 (corr. for H_2)	0.0671 (corr. for H_2)
H_2 injection pressure and temperature (psia/°F)	—	1000/697 1000/853	1000/80
Compression work, (Btu/lb)			
Air	—	339 339	339
Hydrogen	—	4682 5219	100
Mixture or total	444	461 476	332
Air pressure and temperature before mixing (psia/°F)	—	534/1288 534/1288	534/1288
Pressure and temperature after compression and mixing (psia/°F)	529/1233	741/1249 766/1306	615/958
Expansion work after constant volume combustion (Btu/lb)	898	904 907	879
Net work (Btu/lb)	454	443 431	547
Thermal efficiency	0.312	0.305 0.280	0.376
Relative thermal efficiency	1.000	0.976 0.897	1.205
Relative displacement/charge mass	1.000	0.718 0.718	0.718
Relative engine mechanical efficiency	1.000	1.062 1.062	1.057
Relative hp/displacement	1.000	1.444 1.327	1.773
Relative total displacement including compressor or pump	1.000	1.001 1.158	0.6204

Note: Compression ratio = 14:1.

    Inlet air pressure and temperature = 14 psia and 120°F.

    Fuel/air ratio = 0.029:1; ∅ = 1.0.

    Compression and expansion efficiencies = 0.75.

    Two-stage hydrogen compressor with intercooler for 200°F inlet to second stage.

From Donnelly, J. J., Jr., Escher, W. J. D., Greayer, W. C., and Nichols, R. J., *Int. J. Hydrogen Energy,* submitted. With permission.

a four-cylinder engine with 2.3-ℓ displacement. External and internal mixture formation methods were compared using the hardware as shown in Figure 3. External mixture formation was accomplished on the single-cylinder engine using a small flow tube and by means of a propane type carburetor (mixer) for the four-cylinder engine. A semi-internal mixture formation was constructed by injecting hydrogen through ports in the intake valve seat ring in the manner proposed by Adt et al.[26] Direct cylinder injection (internal mixture formation) was tested using a hydraulically controlled valve (Figure 3) which allowed changes to be made in injection timing. Three hydrogen supply pressures were tested: 6, 3, and 0.2 MPa.

TABLE 3

Comparison of Different Mixture Formation Methods

Mixture forma-tion method (base)	Vol efficiency (%)	At same displacement		At same performance	
		Charge heat value (%)	Performance (%)	Displacement (%)	Special fuel consumption (%)
Gasoline engine ($\phi = 1$)	86	100	100	100	100
External mf plus 25% EGR ($\phi = 1$)	72	71	64	158	113
External mf lean operation ($\phi = 0.67$)	86	63	53	195	123
External mf plus water induction ($\phi = 1$)	88	87	83	121	105
Internal mf early (inlet valve open) ($\phi = 1$)	86	85	81	126	107
Internal mf late (inlet valve closed) ($\phi = 1$)	86	121	126	79	95

Note: mf = mixture formation; performance refers to power output.

From Drexl, K. W., Holzt, H. P., and Gutmann, M., paper presented at the NATO/CCMS 4th Int. Symp. Automotive Propulsion Systems, Vol. 2, Session 6, Arlington, Va., April 1977. With permission.

Full load data obtained with the different mixture formation methods are shown in Figure 4. The break mean effective pressure (BMEP) obtained with external mixing depends on the flashback control method. Water induction allows higher BMEP than does EGR because of the oxygen dillution by EGR.

Direct cylinder injection gives the highest BMEP, the value of which depends on the hydrogen supply pressure and injection timing. Figure 5 shows the effect of injection start on BMEP, exhaust temperature, and pressure rise per degree of crank angle. Hydrogen supply pressure for Figure 5 is 6 MPa. Figure 6 gives a theoretical prediction of the combined effect of valve opening area and injection timing for a constant supply pressure of 2 bar (0.2 MPa) on theoretical IMEP.

Low-pressure injection (Figure 6) is a compromise between external mixing and high-pressure internal mixing with late injection. The dependence of IMEP on the valve area is caused by displacement of air by hydrogen because of the early start of injection when the intake valve is still open. The maximum area of the injection valve depends on the space available in the cylinder head. A proposed design of the Daimler Benz 2-MPa injection valve is shown in Figure 7. This valve is designed to fit a cylinder head with 0.6-ℓ displacement per cylinder. Variable lift control is used for port load operation.

Daimler Benz made a comparison of fuel economy of hydrogen and gasoline using the four-cylinder engine with external mixture formation as shown in Figure 8. Hydrogen has better fuel economy at low speed and load because hydrogen combustion more closely approximates constant volume combustion and because of the lean operation with less throttle valve losses. This comparison was made at the same compression

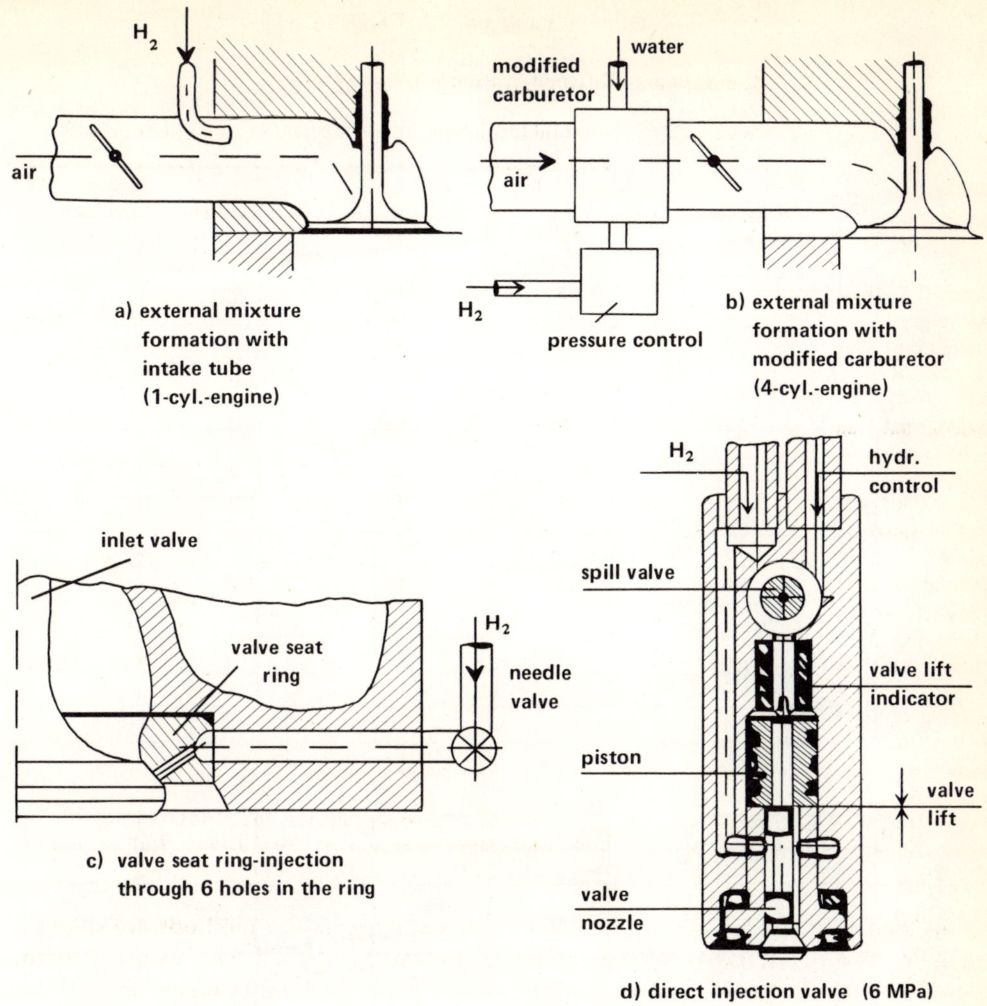

FIGURE 3.   Mixture formation methods tested by Daimler Benz.[29]

ratio (CR) and, therefore, does not indicate the improvement obtained by increasing CR well above the autoignition limit on CR with gasoline. Autoignition has not been observed with hydrogen at CR up to 29.

1.4.3. Emission of NO_x

Emissions of unburned hydrocarbons, carbon monoxide, oxides of sulfur, and carbon dioxide are not observed in hydrogen engines except in trace amounts caused by combustion of oil blow-by. There remains the problem of NO_x (oxides of nitrogen) control. Any air-breathing engine is capable of forming NO_x from the nitrogen in the air. All that is required is that the combustion temperature rise above the threshold for NO_x formation and that the combustion products cool more rapidly than the dissociation reaction rate back to N_2 and O_2 (frozen equilibrium).

A hydrogen engine is capable of higher reaction temperature (and energy for a given mass of air) and higher flame speeds. It is expected, then, that a hydrogen engine would produce more NO_x. While this is certainly true, the control of NO_x formation has been demonstrated by several techniques. In the case of the external mixing design,

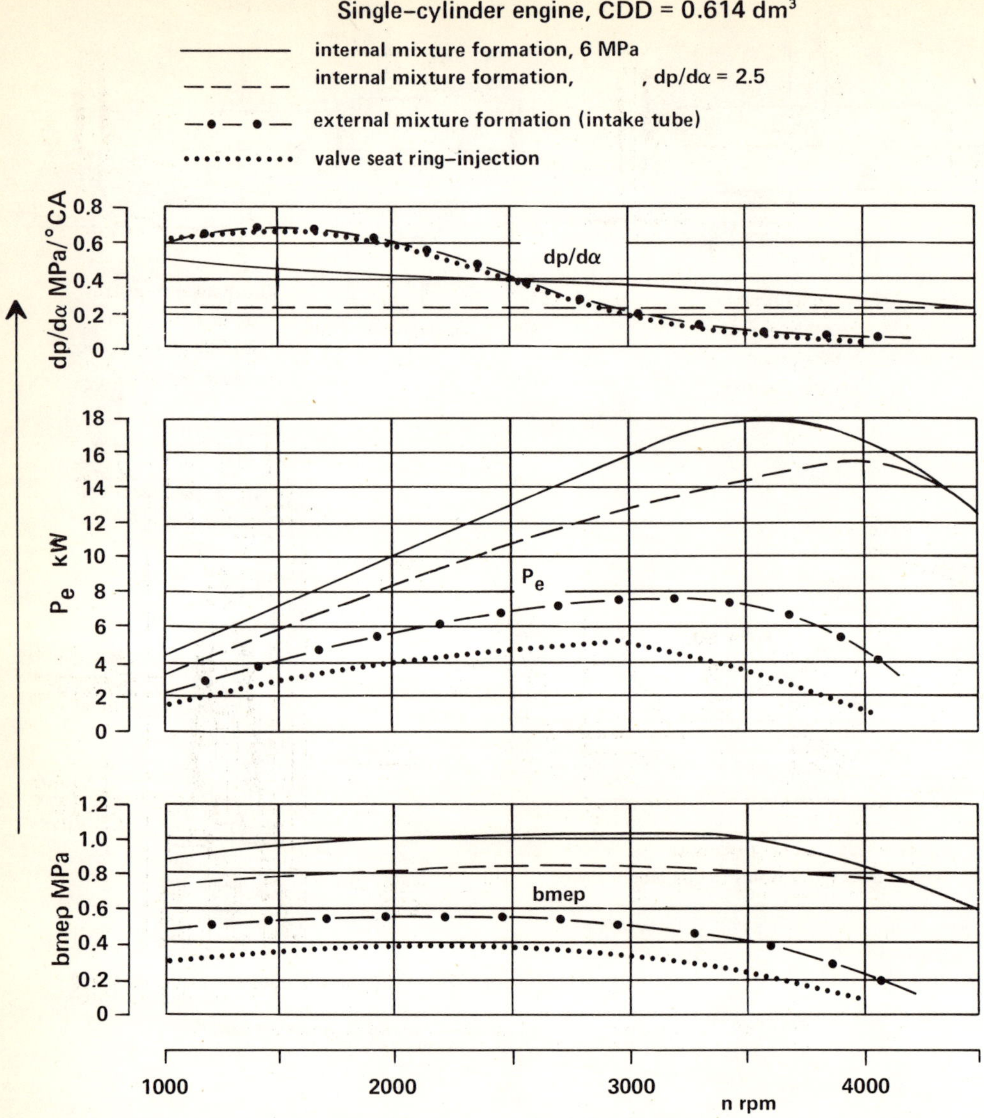

FIGURE 4.   Data at wide-open throttle.[29]

the control of NO_x takes place simultaneously with flashback control and is therefore necessary if reasonable power levels are to be achieved. NO_x control methods parallel preignition and flashback control methods discussed previously; ultra lean combustion, low load, water induction, EGR, and timed combustion of a direct cylinder injection system.

Ignition timing has been found to be an important parameter in NO_x formation since excessive ignition advance causes higher peak temperature. Spark timing is essentially a function of equivalence ratio and CR. It is only weakly a function of revolutions per minute (rpm) and manifold vacuum. It was shown by Burstall[32] in 1927 that less ignition advance was required for hydrogen than for hydrocarbons because of higher flame speed. The change in timing with equivalence ratio and CR is shown in Figure 9.

Obtaining the proper timing is a more difficult task for very lean mixtures. It has

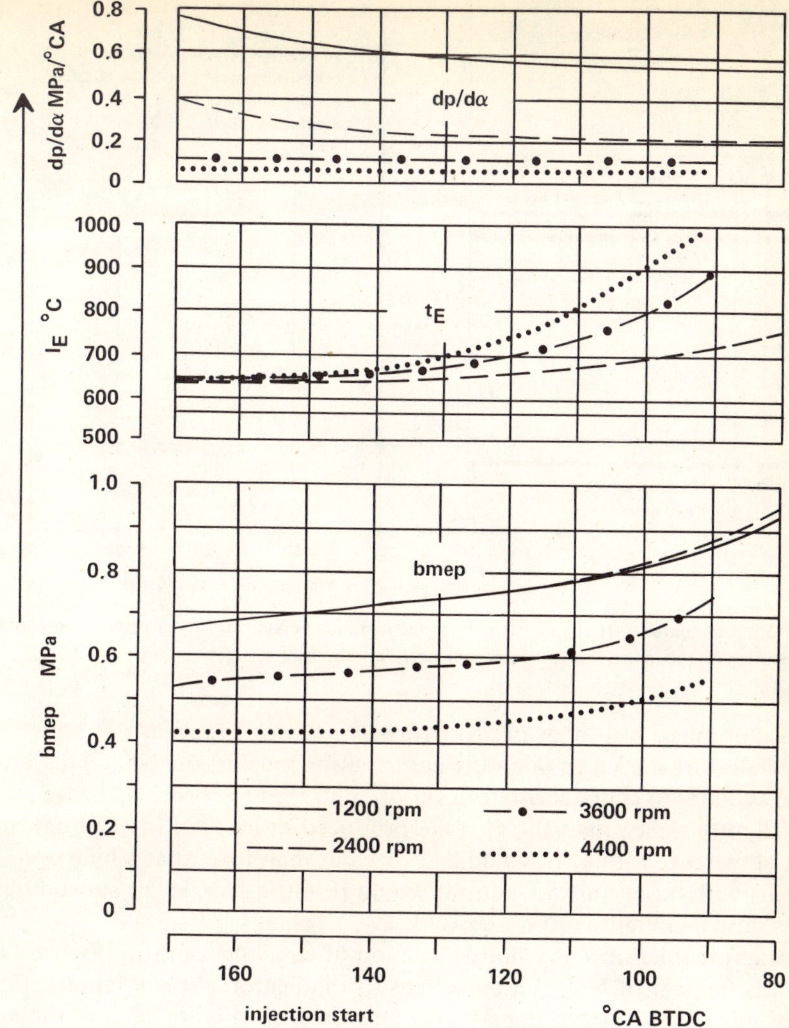

FIGURE 5.   Influence of injection start on BMEP, exhaust gas temperature, and pressure rise.[29]

been recommended that the setting be selected using an NO_x instrument in combination with the conventional torque and fuel flow methods.[34] Both torque and efficiency peaks occur at nearly the same advance with the efficiency peak slightly on the retard side of the torque peak. NO_x increases exponentially with advance. Therefore, the recommended method for timing the engine with a fixed equivalence ratio is to find the torque peak and then retard the timing while watching both torque (at constant rpm) and NO_x level. Fuel flow rate should also be monitored. A compromise must be made between NO_x level and reduced torque. Both the torque and efficiency peaks are well rounded. Therefore, a substantial reduction in NO_x is possible for a small sacrifice in power and efficiency.[34]

Equivalence ratio is a dominant parameter in NO_x formation because of energy flow and temperature rise in the engine. Figures 10 and 11 show an exponential increase in NO_x with increasing ϕ.[34] Water induction seems to dramatically reduce the level of NO_x at all engine rpm but does not change the exponential character of the increase with ϕ

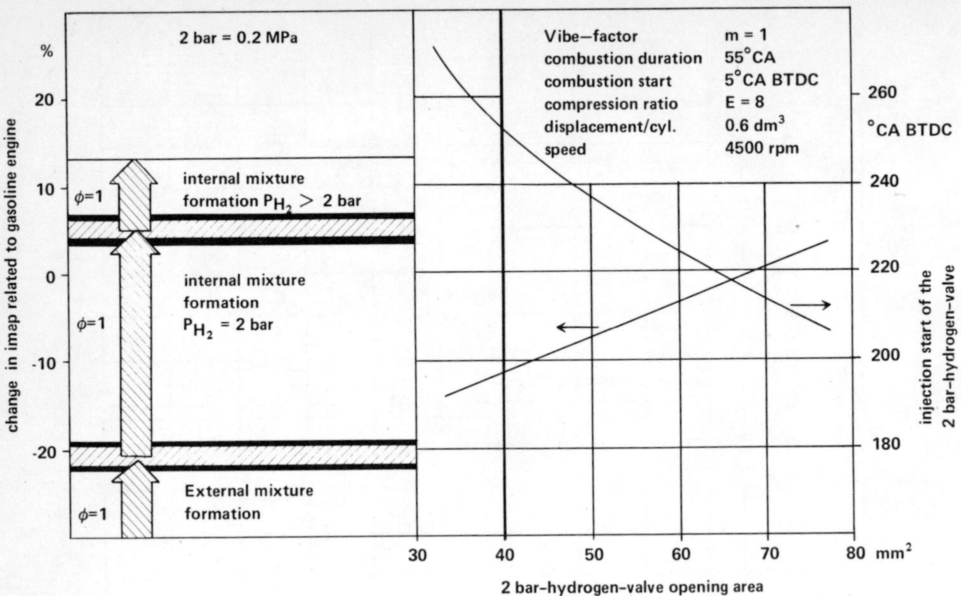

FIGURE 6.   Calculated IMEP values for hydrogen engines related to the gasoline engine and injection start depending on the hydrogen-valve opening area for a hydrogen pressure supply of 2 bar.[29]

for a constant mass ratio of water to hydrogen. It was found in this study[34] that one or two cylinders of the V8 engine were contributing most of the NO_x. This was because of a slight cylinder to cylinder mixture variation in ϕ on the order of 10%. The cylinder that was slightly richer than the average produced more NO_x in accordance with the exponential increase with ϕ. It would be expected, therefore, that a hydrogen-air mixer (carburetor) with more uniform output would result in NO_x levels substantially below the curves shown in Figures 10, 11, and 13.

Exhaust gas temperature is a linear function of equivalence ratio (Figure 12). Therefore, a measurement of NO_x is a more sensitive indicator of peak temperature change than is exhaust temperature. It may also be seen from Figure 12 that the amount of inducted water changes the exhaust temperature only slightly for water to hydrogen mass ratios of interest for NO_x control. This conclusion also held for engine efficiency and power, which were found to be essentially independent of water induction level.[34]

NO_x decreases exponentially with increased water induction as shown in Figure 13. Again, the NO_x level shown is the average for the eight cylinders and, therefore, is higher than could be obtained with more uniform mixing of hydrogen and air. Water induction is an effective means of controlling NO_x without loss of power, efficiency, or exhaust temperature. The effectiveness of water induction increases with rpm.

This remedy is not without problems. Induction of water into the cylinder increases the amount of water getting into the engine oil. Higher block temperatures reduce the amount of water going past the rings. The problem may also be dealt with by lowering the pressure on the oil pan and keeping the oil hot.[35]

The point at which the water is added to the gas stream also has an effect on the amount of NO_x produced.[23] Since NO_x increases exponentially with ϕ, slight mixture nonuniformities within the cylinder (as with hydrogen injection), as well as cylinder to cylinder variations, will increase NO_x as noted previously. It is important, therefore, to have the NO_x retardant (water) nonuniform in the same manner. Perhaps the best way to achieve this, although not always the most convenient, is to induct the water

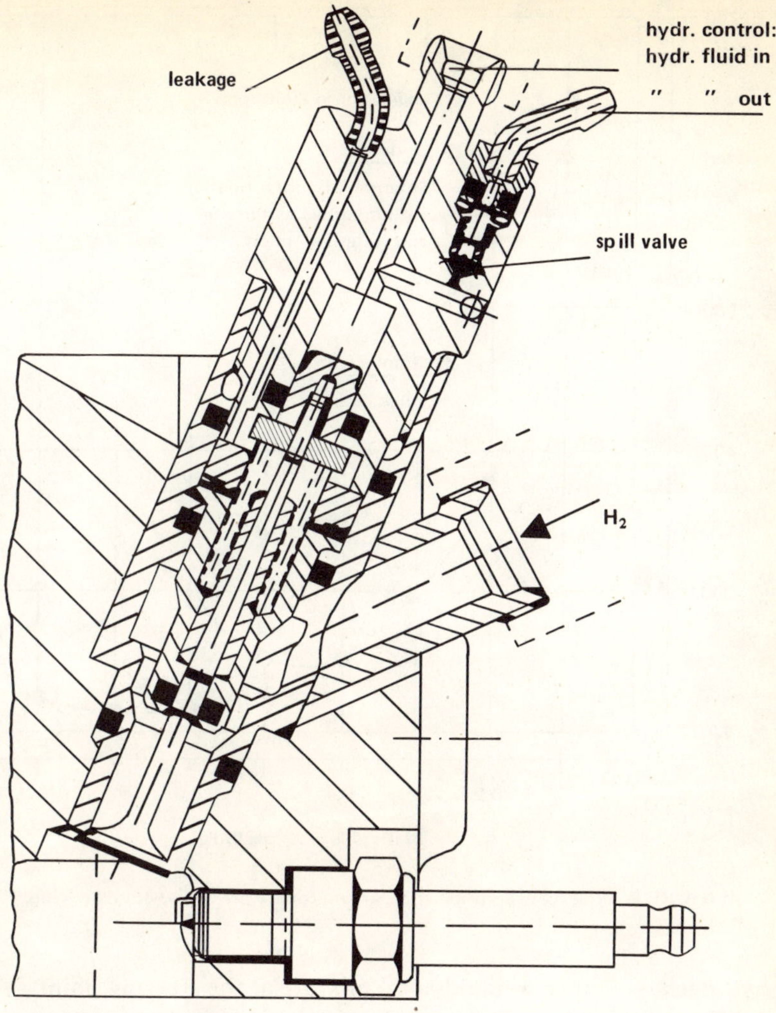

FIGURE 7. Hydrogen injection valve design for internal mixture formation. Design supply pressure is 2 bar.[29]

into the hydrogen stream prior to mixing with the air. When this was done, a reduction was observed in NO_x by an order of magnitude for the same amount of inducted water.[23]

1.5. SPECIAL CONSIDERATIONS FOR A HYDROGEN VEHICLE

1.5.1. Hydrogen Storage

The low energy density of hydrogen (on a volume basis) has been noted and is clear from Table 1. The fact that this implies a considerably greater storage volume than would be required for gasoline is not so much the difficulty, however, as is the problem of obtaining and maintaining hydrogen in a suitably dense form. Several methods have been suggested: as a high-pressure gas, as a metallic hydride, as a cryogenic liquid, as "slush" hydrogen, and chemically bonded. "Chemically bonded" implies a chemical rich in hydrogen that would be decomposed to yield hydrogen as necessary. Hydrogen might be generated from the catalytic decomposition of methanol, for example.

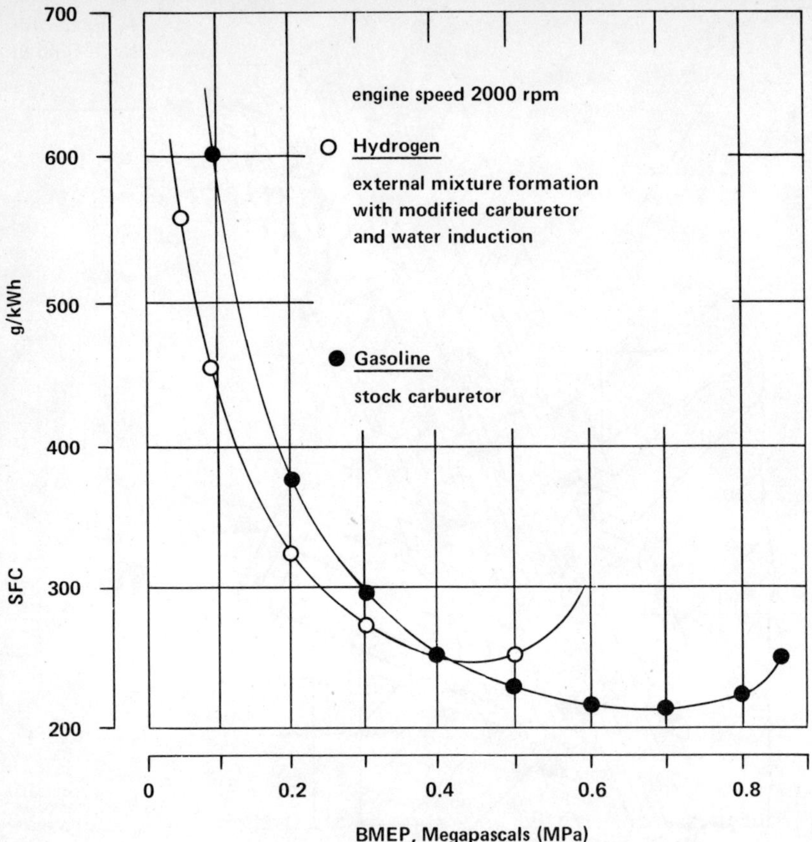

FIGURE 8.   Fuel consumption comparison for a 2.3-ℓ four-cylinder engine.[29]

"Slush" hydrogen — that is, a liquid-solid mixture at the freezing point — was earlier thought to offer promise as a maximum density state of hydrogen. More recent experience, however, has indicated that it is less than practical. Storage as a gas seems clearly impractical. Extrapolating the figures available from the UCLA hydrogen car,[21] excessively high pressures and a severe weight penalty would be incurred. The feasible alternatives for storage of hydrogen as such, therefore, reduce to two — as a metallic hydride and as a liquid. The overall problem has been reviewed by Williams[36] who concluded that liquid storage offered the only practical alternative. Since then (1973), however, continued research and development in the area of hydrides now make the hydride system one of feasibility comparable with the liquid.

Reilly and Wiswall[37] first reported that certain metals and metallic compounds have the property of absorbing hydrogen with the release of heat and subsequently regenerating it with the addition of heat. Hoffman et al. suggested the use of this phenomenon as the basis of a hydrogen storage system for vehicular use.[38]

The magnitude of the heat effect, the temperature-pressure equilibrium relation, and the rate at which each process (absorption and regeneration) occurs are all critically important properties of the system. Rather qualitatively, the desirable properties of the hydride may be listed as:

1.   Low heat of dissociation

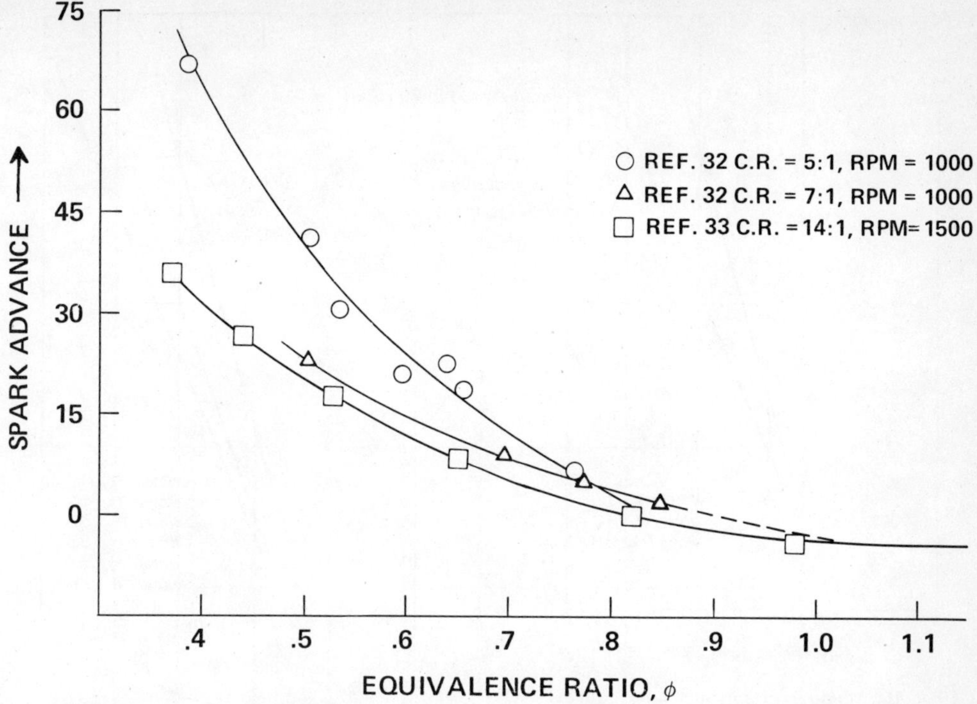

FIGURE 9.   Ignition timing as a function of equivalence ratio and compression ratio.

2. Adequate pressure discharge level (several atmospheres) at a comparatively low temperature (60 to 100°F); an idealized representation of the pressure-composition-temperature relation is shown in Figure 3
3. High percentage (by weight) of hydrogen carried
4. Fast release rate
5. Minimal pulverization on continued use

Unfortunately, no hydride known at the present time has an entirely desirable "mix" of these properties. A few examples will illustrate the problem (Figure 14).

Iron-titanium ($FeTiH_{1.6}$) has a relatively low heat of dissociation (7250 Btu per pound of hydrogen), and dissociation occurs at acceptably low temperatures (100°F) with a satisfactory pressure level. Experimental results indicate a satisfactorily high release rate. This combination would permit supplying all the heat for releasing the hydrogen required from either the engine exhaust or from the engine coolant system, or a combination of both. The "hydrogen capacity" of iron-titanium is low, however, and the bulk density high. Based on the data of an experimental storage reservoir developed by the Brookhaven group,[40] approximately 1900 lb of the hydride would be required to fuel an automobile with a cruising range of 300 mi. The volume required would be approximately 12 ft³, or 90 gal. These figures are for the hydride only — exclusive of the containment vessel, heat exchangers, and other equipment required.

Magnesium-based hydrides offer the advantage of a higher hydrogen storage. Capacities of 6 to 7% have been reported, which would imply a total weight of approximately 425 lb, based on the 300-mi cruising range postulated above. The necessary volume would be in the neighborhood of 7.5 ft³ or 57 gal. Again, these figures are for the hydride only. However, the heat of dissociation is high — on the order of 16,000

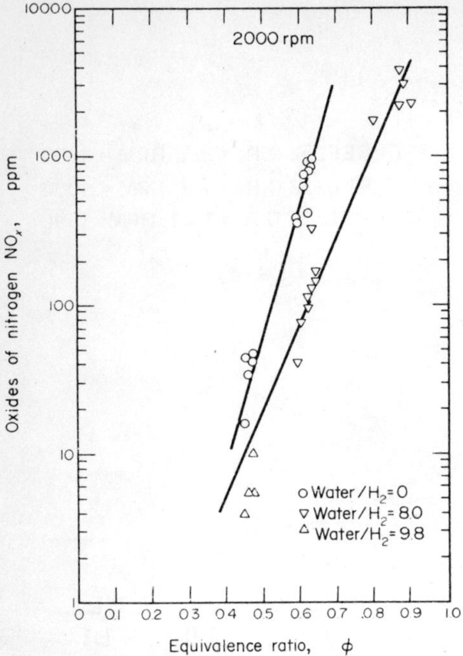

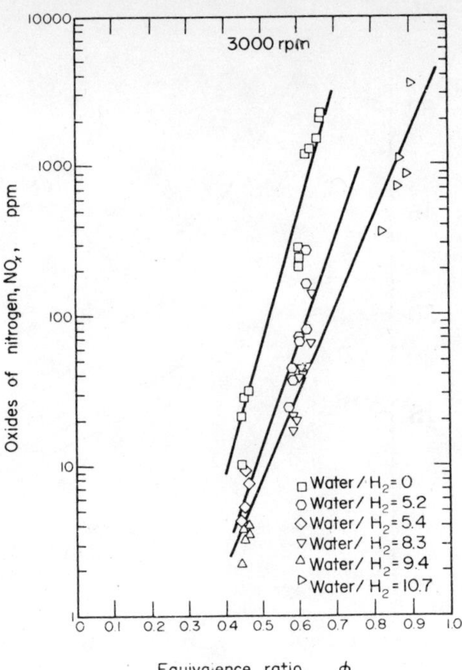

FIGURE 10.   Oxides of nitrogen and equivalence ratio measured at each exhaust valve (2000 rpm). (From Woolley, R. L. and Hendrickson, D. L., *Int. J. Hydrogen Energy*, 1, 401, 1977. With permission.)

FIGURE 11.   Oxides of nitrogen vs. equivalence ratio measured at each exhaust valve (3000 rpm). (From Woolley, R. L. and Hendrickson, D. L., *Int. J. Hydrogen Energy*, 1, 401, 1977. With permission.)

Btu per pound of hydrogen. While pressure levels are satisfactory (1 to 10 atm), the corresponding equilibrium temperatures are comparatively low with hydrogen-fueled engines, and neither the energy of the engine exhaust nor the cooling water system could be used to supply the heat necessary to release the hydrogen: the energy available (above 650°F) is well below that required. (The situation is depicted in Figure 15.) Therefore, auxiliary energy must be used in an amount roughly equivalent to 15% of the energy supplied to the engine. This must, of course, be charged to the engine and would be a severe performance penalty. In addition, there seems to be some tendency of magnesium-based hydrides to continue to fractionate with usage.[41]

The alternate strategy to selection of either all magnesium-based hydrides or all titanium-based hydrides is to use a combination of the two. The magnesium hydride container is sized to match with the amount of energy that can be extracted from the exhaust at the necessary temperature. The remaining hydrogen storage is contained in titanium-iron. Obviously, there exists a significant engineering design problem to provide for adequate heat transfer in a limited space. A prototype design by Daimler Benz that is based on the dual hydride concept is shown in the section of the chapter on prototypes.

While there is no known operating experience with what might be called "onboard generation of hydrogen" systems, it is appropriate to discuss briefly some possibilities. Breshears et al.[42] have investigated a scheme falling in this category at great length. The process was initially suggested primarily to reduce pollutant emission. A fraction of the gasoline is diverted from the flow path to the engine and is passed through a catalytic converter and decomposed to yield hydrogen. The hydrogen is then injected

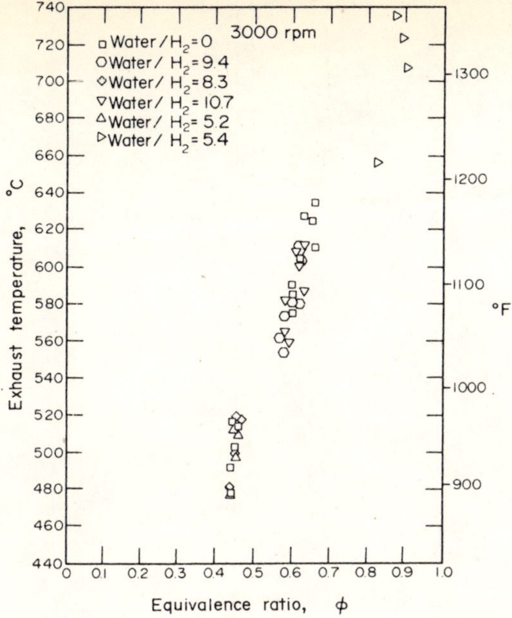

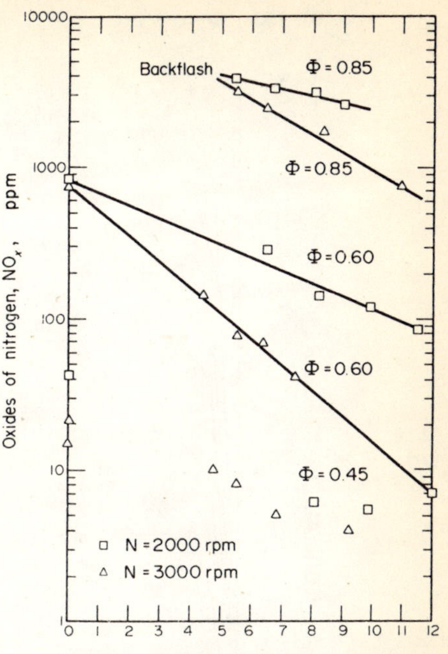

FIGURE 12.   Temperature of manifold exhaust gas at exhaust valve. (From Woolley, R. L. and Hendrickson, D. L., *Int. J. Hydrogen Energy*, 1, 401, 1977. With permission.)

FIGURE 13.   Effect of water, equivalence ratio, and engine speed on NO_x formation. (From Woolley, R. L. and Hendrickson, D. L., *Int. J. Hydrogen Energy*, 1, 401, 1977. With permission.)

into the engine with gasoline, thus allowing the engine to run at a substantially lower equivalence ratio with a corresponding decrease in the pollutants emitted. While this would not be practical to generate hydrogen as the sole fuel fed to the engine, it is illustrative of the general idea.

Kester et al.[43] have suggested that methyl alcohol might be reformed with steam to yield hydrogen which would be the actual engine fuel. Thus

$$CH_3OH \pm H_2O \rightarrow 3H_2 + CO_2$$

The CO_2 would not need to be removed since it would serve as an appropriate diluent. Alternatively, Ullman and Van Vorst have suggested[44] catalytic decomposition as follows:

$$CH_3OH \xrightarrow{cat} 2H_2 + CO$$

There would appear to be a slight energy advantage in the catalytic decomposition as compared to the steam reformation.

Ullman and Van Vorst[44] also proposed a sodium borohydride-water reaction to generate hydrogen:

$$NaBH_4 + 4H_2O \rightarrow NaB(OH)_4 + 4H_2$$

This system has been shown to work well in the laboratory and has a high effective density of hydrogen. The foreseeable economic considerations are so unfavorable, however, that it can be considered only for very specialized application.

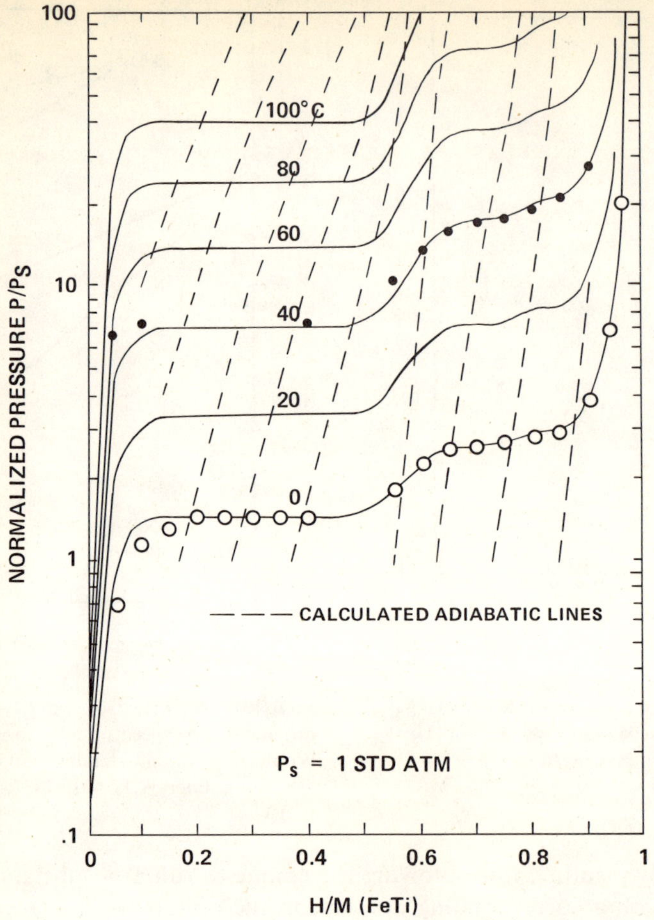

FIGURE 14.   Model of dissociation isotherms and adiabatic lines for
TiFe hydride (Reference 39 curve fit based on Reference 37 data).

1.5.2. Range

Vehicle range is limited by the rather low weight percent of the hydride, weight-carrying capacity of the vehicle, the available volume for placing hydrides in the vehicle, the vehicle fuel efficiency with the added weight, and the recharge time allowed. A similar set of constraints exists for liquid hydrogen storage, although the effect of the added weight is much smaller. A set of equations to estimate the range is given below. These equations were used to evaluate design trade-offs in conversion to hydrogen of a 19-passenger bus for the city of Riverside, California by the Billings Energy Corporation.[45]

$$P = (D_{RR} + D_{WR}) V/E_{dt} \qquad (17)$$

$$D_{RR} = (M_V + M_{MH}) f(V) \qquad (18)$$

$$D_{WR} = \tfrac{1}{2}\rho_{air} V^2 C_D A_f \qquad (19)$$

$$\dot{m}_{H_2} = P/(e \, \Delta h) \qquad (20)$$

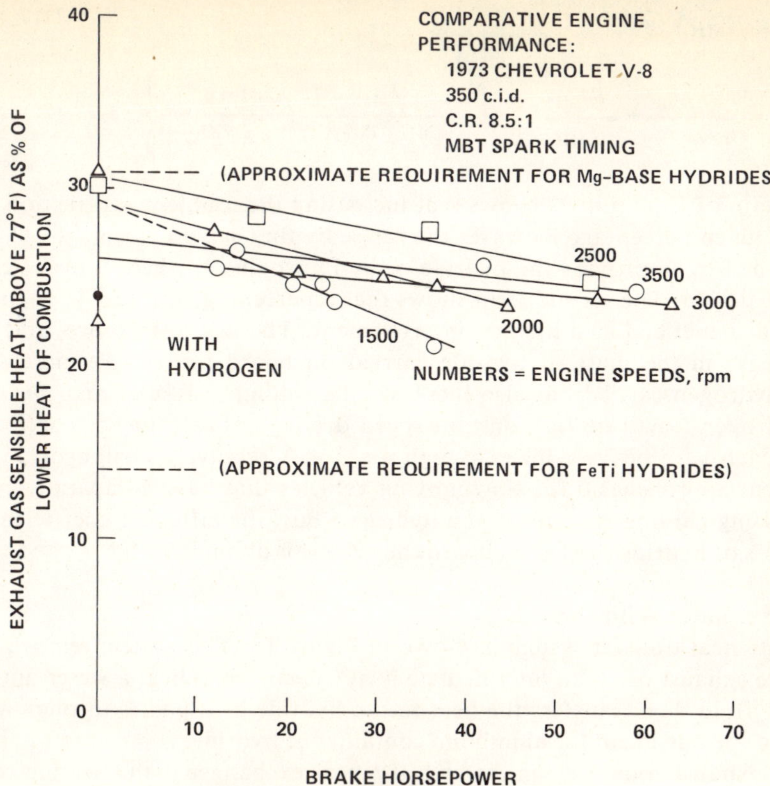

FIGURE 15.   Sensible heat of exhaust gas.[27]

Equations 17 to 20 were used to calculate the rate of hydrogen flow. There are several formulas available for the velocity function in Equation 18. They may be found in textbooks on internal combustion engines. Actual measurement of the vehicle drag may be obtained from coast-down tests.[46]

The maximum amount of weight that can be carried is given by the difference between the gross vehicle weight (GVW) and the curb weight plus passenger and driver. The amount of hydride is also limited by the space available for hydride tanks. In the case of the Riverside hydrogen bus, a special-order bus was selected to make these two parameters as large as possible. An Argosy 24-ft bus was ordered to be constructed on the heavy duty chassis normally used for the 26-ft version. The seating arrangement for 21 passengers was then altered by removing one bench seat up front, giving a 19-passenger capacity. The 150 lb (68 kg) per passenger that was saved was used to carry more hydride. This left a weight carrying capacity sufficient to hold 2000 lb (907 kg) of hydride plus the weight of the aluminum containers. The amount of stored hydrogen then depended on the usable weight percent of the hydride, provided that sufficient space could be found to carry the tanks.

The usable weight percent for TiFe is somewhere between 1.0 and 1.9, depending on recharge conditions of discharge pressure, flow rate, and heat transfer. The range at 55 mph (88 km/hr) was thus estimated to be between 80 and 170 mi (129 and 247 km). Design choices determine the final result obtained.

The relative importance of design decisions on the range can be estimated by writing an equation for vehicle range at constant speed and then using logarithmic differentiation to obtain influence coefficients.

$$R = m_f M_{MH} V / \dot{m} \tag{21}$$

$$\frac{dR}{R} = \frac{dm_f}{m_f} + \frac{de}{e} + \frac{dM_{MH}}{M_{MH}} \left\{ 1 - \frac{M_{MH}}{(M_V + M_{MH})} \frac{D_{RR}}{(D_{RR} + D_{WR})} \right\} \tag{22}$$

The first term of Equation 22 shows that increasing the usable mass fraction of the hydride by a given percentage increases the range by that same percentage. This may be accomplished by improving the hydride or by improving the heat transfer characteristic of the design. The second term shows that a percentage increase in engine performance is as beneficial as a hydride improvement. The last term shows that a percentage increase in the mass of hydride carried on board the vehicle increases the amount of hydrogen carried but also increases the rolling resistance and, hence, the required hydrogen flow rate for constant speed driving. This influence coefficient is seen to approach 1 for heavy vehicles at high speed with relatively small hydride tanks. The coefficient approaches 0 for slow-moving vehicles that have comparatively large fuel tanks. Using parameters typical of a hydrogen bus, the influence coefficient for a change in mass of hydride carried on board the vehicle is about 0.9.

1.5.3. Heat Exchange With Hydride

A candidate heat transfer system is shown in Figure 16. This system removes waste heat from the exhaust using an intermediate heat exchanger to heat a water-antifreeze mixture. The fluid then transfers the heat to the hydride as it passes through a water jacket around the outside of the aluminum container of hydride.

All of the exhaust could be run through the heat exchanger if the heating requirement for the hydride were great enough. Such is sometimes the case if the hydride tanks are not insulated. One method to avoid boiling the fluid is to divert a fraction of the exhaust. This method also permits a greater reduction of temperature in the exhaust fraction passing through the heat exchanger so that some water may be condensed and collected for reuse in the engine.

An important characteristic of the heat transfer system for a dual fuel vehicle is that the hydride must not be heated while the vehicle operates on the alternate fuel. This may be accomplished by closing the diversion valve in this design.

Although attractive in principle, this system has been found to be troublesome because of poor reliability of the diversion valve.[45] Dependable operation of any moving part in the exhaust stream is a significant design problem in and of itself. Another disadvantage of this system is the increased back pressure on the engine. Back pressure is an important consideration for engine power, efficiency, and tendency to flashback.

A second heat transfer system is shown in Figure 17. This design uses engine coolant as the heat transfer fluid, the engine block becomes the heat exchanger, and the water jacket around an aluminum tank is again used. There is sufficient energy in this system to release the hydride. Roughly 35% of the fuel lower heating value is transferred to the engine coolant as waste heat. Only 13% of the hydrogen lower heating value is required to release the hydrogen from titanium-iron hydride. Losses to the wind from lines and tanks increases the heat requirement such that, in practice, the radiator does very little cooling.

The system in Figure 17 is very reliable and has been found to perform adequately, although some supplementary heating from the exhaust may be advantageous.[45] When operating on the alternate fuel, the pump in the circuit is switched off, and fluid circulation is reduced below the level of heat losses to the surrounding air. One disadvantage of this system is that a means is not provided for water condensation from the exhaust, and a water storage tank must be added.

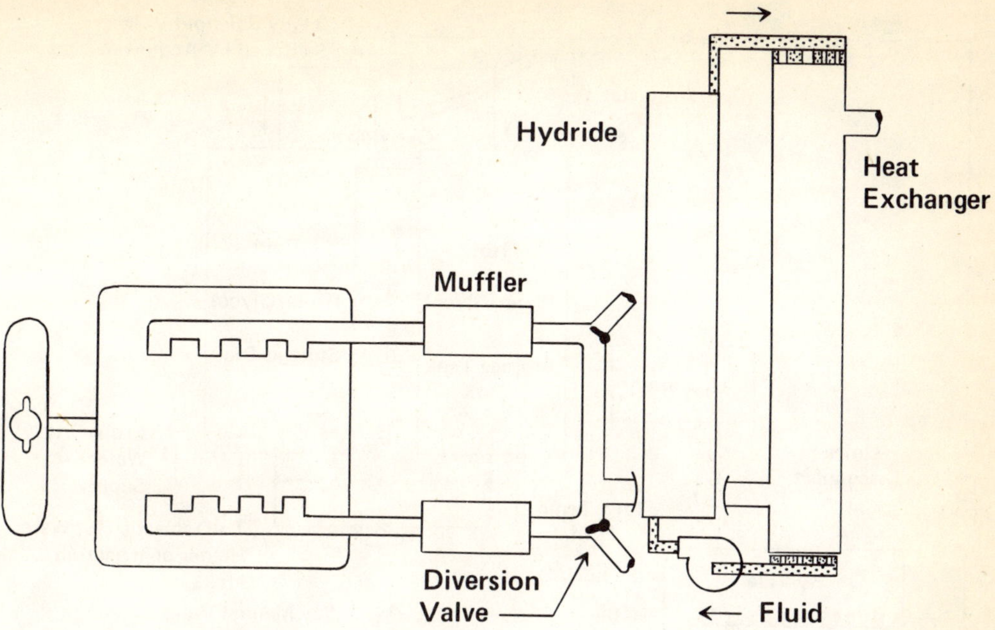

FIGURE 16.   Hydride heating circuit using engine exhaust heat.[45]

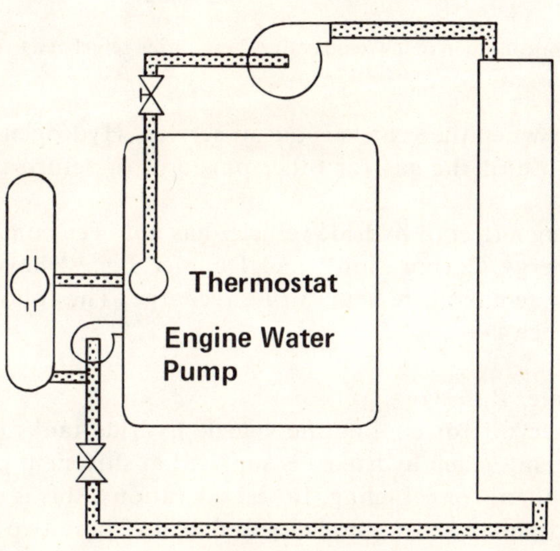

FIGURE 17.   Hydride heating circuit using waste heat
from the engine coolant.[45]

1.5.4. Refueling a Hydrogen Vehicle

The refueling of a liquid hydrogen vehicle is done in a manner analogous to that used in the space program. A refueling station would service a fleet of vehicles — such as a transit bus fleet or a fleet of support vehicles at an airport that has hydrogen-fueled aircraft. It is expected that the refueling system would have a means provided for recovery of "flash-off" hydrogen that is released during vessel chill down and

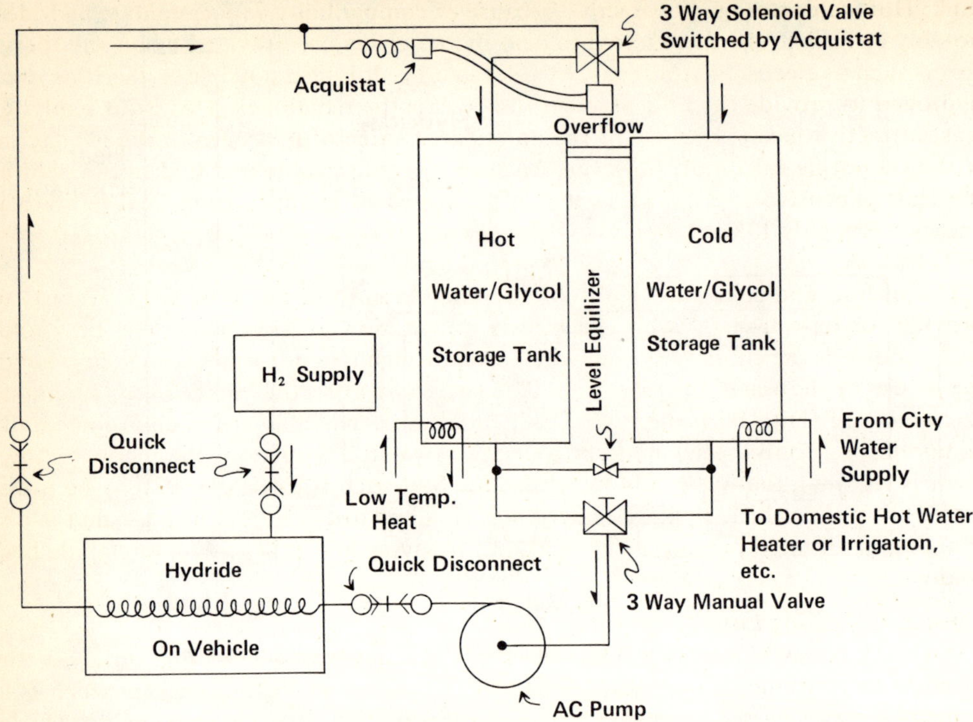

FIGURE 18.   System for collection of low-temperature heat during refuel of a hydrogen vehicle.[47]

some "boil-off" losses when the vehicle is out of service. Hydrogen recovery could be either in the form of use of the gas for other purposes or return to the liquefaction plant if possible.

A system for refueling a fleet of hydride vehicles has not been constructed. Refueling studies by Billings Energy Corporation[47] and Daimler Benz[48] have shown that it is possible to reclaim low temperature heat during recharge. The following material was excerpted from Reference 47.

1.5.4.1. Hot/Cold-Water Reservoirs

A practical cold reservoir for cooling the vehicle hydride tanks during refueling is the culinary water system. When hydrogen is supplied at sufficient pressure, this is the only heat reservoir required for refueling. In test operations, this is the method chosen for cooling the hydride because of simplicity and cost. These two factors still apply for a fleet application. The only addition would be provision for using the water for some constructive purpose after it has been warmed by the hydride.

In high-temperature vehicle hydride designs where the cooling water fills the engine exhaust passage, the water would be contaminated and useable only for irrigation. For low-temperature hydride vehicles, it is an easy matter to design a system of heat exchange that does not contaminate the water. It then becomes possible to use the culinary water as input to a hot-water system and save some of the energy released by the hydride. A flow-through system is one possibility. A more elegant system that saves the low-grade heat at a higher, and therefore more useful, temperature is shown in Figure 18.

Systems that have hydride storage at the station require a heat source at a high enough temperature to build sufficient hydrogen pressure for flow into the vehicle tank. Hot-water storage is attractive because of compatibility with candidate hydrides, notably FeTi. The energy required to heat the water may be provided by on-site energy forms such as electric, natural gas, or solar energy. It is also possible to sacrifice some hydrogen to provide the heat necessary to release the remainder. This is probably the least attractive means because of the anticipated value of hydrogen in the near term. It should not be ruled out, however, because it is conceivable that hydrogen could be the least expensive fuel available if it is generated in quantity from coal or nuclear energy.* Roughly 13% of the lower heating value is required to release hydrogen from FeTi hydride.

Use of hot- and cold-water reservoirs is considered to be the most attractive way to provide for heat transfer with the vehicle and station hydride units. The best cold-water source is the culinary system with flow-through to a hot-water system in a building or use of the water for irrigation. The two most attractive methods for obtaining hot water are (1) to burn the lowest cost fuel that is available and is environmentally compatible or (2) to use solar energy to heat the water. Fortunately, the energy requirement is less than that necessary to release the hydrogen because energy may be transferred from the vehicle hydride undergoing recharge to the station hydride that is discharging at elevated pressure. The method for saving and transferring low-grade heat follows.

1.5.4.2. Collecting Low-Temperature Heat

An alloy releases heat as it hydrides. Such heat must be removed in order for the reaction to continue to the maximum value. Consider an FeTi system in which cold culinary water is passed through a heat exchanger with the hydride vessel in the vehicle. During the early portion of the refueling period, the water may be warmed to boiling temperature because the equilibrium temperature at pressure is in that range. The flow rate of water can be adjusted for an outlet temperature less than the hydride equilibrium temperature if desired. It is easy to visualize a system where the water flow rate is regulated to maintain a preset temperature level. The problem with such a system is that there will be no flow and, hence, no cooling as the hydride cools below the desired water outlet temperature. Additional cooling is often desirable to achieve the highest uptake of hydrogen. These tests have revealed that, over a large portion of the refueling period, for a complete refueling, the outlet water temperature is much less than the maximum achieved earlier in the period. Hence, if cold water runs continuously past the hydride, the bulk temperature of all the water is low. If water is circulated from a cold tank to a "hot tank", the final minutes of the refueling process serve to dilute the temperature of the hot tank. If water is circulated back to the cold tank, the temperature of that vessel increases, thereby reducing the heat transfer rate, and the released heat is again at a temperature only slightly above the original (depending on tank volume).

A method of saving low-grade heat at a temperature near the maximum is shown in Figure 18. In this system, water — or a water/glycol mixture — is stored in two tanks. Flow from one of the tanks to the vehicle is selected by the position of a three-way valve. When recharging begins, cold fluid is selected. This fluid is heated by the reaction taking place in the hydride. The temperature of the returning fluid is sensed by an aquastat or similar device. When the temperature exceeds a preset value, the solenoid valve automatically opens the return path to the "hot" tank; when the temperature is below this value, the flow returns to the "cold" tank.

---

*   In systems that use hydrides requiring temperatures above the normal boiling point of water (none is recommended at present), it is quite possible that hydrogen (or alternate fuel) would be burned to supply the heat. No energy storage would be required other than the fuel itself.

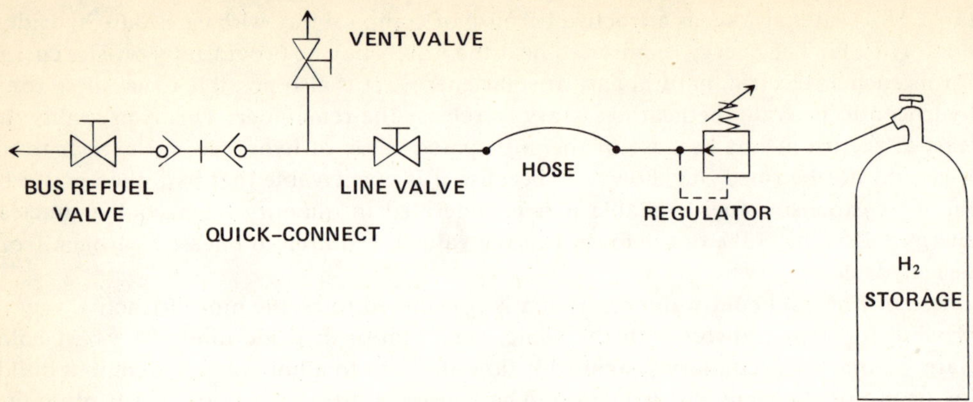

FIGURE 19.   Hydrogen refuel line.[47]

The objective of this arrangement is threefold: (1) the "cold" storage tank is not heated by return of "hot fluid" above a certain temperature, (2) hot water above a certain temperature is collected for other uses, and (3) some hot fluid may be returned to the vehicle just before it leaves.

The temperature of the "hot" water is between 40 and 80°C and would find application in the same designs that low-temperature solar heat is used, such as water heating, building heat, etc. The system could be used to supplement a solar heat design since the heat is generated whenever a vehicle refuels. The heat may also be used to warm a storage hydride system that provides hydrogen to the vehicle.

As the refueling nears completion, the return flow is heated only slightly. Before disconnection, it is possible to flow some fluid from the "hot" reservoir back into the vehicle to warm the tanks. This raises the pressure in the tanks and allows a faster discharge of hydrogen from the vehicle tanks during initial operation. This strategy may also be used to heat a vehicle tank that has become cold because it has been out of service during cold weather.

The cold reservoir must be cooled by some means. In arid regions of the country, a cooling tower may be used to drop the temperature about 20°C from the air temperature. In most locations, the city water supply is sufficiently cold. It may be circulated through a coil in the reservoir as shown and then input to a hot-water heater or used for irrigation.

A typical refueling line connection is shown in Figure 19. The line provides for venting of any air present in the line in order to avoid contamination of the hydride.

1.5.4.3. Increased Thermal Capacity

During the recharge of the Billings hydrogen bus,[44] it was found that 40% of a 3-hr recharge was completed in the first 5 min, 55% after 10 min, 78% after 30 min, and 86% after 1 hr. The rapid initial flow takes place because the heat generated by the hydriding reaction increases the temperature of the hydride, tank, and heat exchanger fluid according to the heat capacity of the system. It is advantageous, therefore, to pull into the refueling station with a cold hydride. Chill-down will take place if the withdrawal pressure is sufficiently low.

The remainder of the refuel period is limited by heat transfer. During the recharge after the initial rapid flow, it is necessary to remove heat from the hydride. The amount of heat removed depends on the hydrogen-charging pressure, the initial temperature of the hydride, and the thermal capacity of the hydride and container. The thermal capacity of the system is inadequate to hold all of the energy unless another material

TABLE 4

Thermal Properties of Inert Materials

	ϱ (lb/ft³)	c (BTU/lb-F)	ϱc (BTU/ft³-F)	k (BTU/hr-ft-F)
Aluminum (pure)	169	0.215	36	145
Copper (pure)	560	0.092	52	227
Iron (pure)	491	0.108	53	44
Lead (pure)	708	0.031	22	20
Magnesium (pure)	108	0.250	27	92
Al 6061-T6	169	0.215	36	97
Bronze (75%Cu, 25%Sn)	541	0.082	44	15
Brass (85%Cu, 9%Sn, 6%fn)	544	0.092	50	35
Stainless steel (18%Cn, 8%N₂)	488	0.11	54	9.4
Sandstone	140	0.17	24	0.94—1.2
Glass	169	0.20	34	0.44
Concrete	119—144	0.21	28	0.47—0.81
Water	62.4	1.00	62	0.345
Ethylene glycol	69.7	0.569	40	0.144
Engine oil	55.5	0.45	25	0.084
FeTi hydride	406	0.15	61	1.0

is added to the hydride and container. For a vehicle design, the best material is one with a high heat capacity in order to minimize the weight addition. For a stationary hydride system, there is a trade-off between minimizing volume (large pc) and mass (large C). An increase in volume results in a larger pressure vessel and more expense. The thermal conductivity of the inert material would also be a consideration.

From Table 4, we conclude that, for lightweight systems, aluminum is a good choice for the container, and water is a good material to increase the thermal capacity. For a low-volume system that may be heavy, steel, stainless steel, or aluminum can be selected for the container while iron, copper, water, and ethylene glycol are good thermal storage materials. The table also illustrates that, for stationary systems, an over design on the amount of FeTi can function effectively for thermal storage; however, this is expensive.

Consider a system in which aluminum tanks are selected to hold FeTi hydride. What weight of water is required to increase the thermal capacity such that the hydride can be charged to 1.6% by weight without exchanging heat with the surroundings? Assume that the aluminum container weighs 27% of the hydride weight (typical) and that the temperature of the total system increases 150°F (83°C) from 40 to 190°F (4.4 to 88°C).

The amount of water required per mass of metal hydride, M_{MH}, for an adiabatic system is calculated as follows (Table 5):

$$\frac{M_{H_2O}}{M_{MH}} = \left[\frac{M_{H_2}}{M_{MH}} \frac{hsg}{T} - C_{MH} - \frac{M_{Al}}{M_{MH}} C_{Al} \right] / C_{H_2O} \qquad (23)$$

$$= \left[\frac{(0.016)(6500)}{150} - 0.15 - 0.28(0.21) \right] / 1.0 = 48.3\%$$

1.6. PROTOTYPE HYDROGEN VEHICLES

1.6.1. Prototypes Using Liquid Hydrogen

The essential difficulty in using a liquid hydrogen system may be summarized rather tersely by noting that its boiling point at atmospheric pressure is 20 K, or −253°C

TABLE 5

Requirements for Adiabatic System

	FeTi hydride	Al tank	Water	Hydrogen
% of hydride	100	28.0	48.3	1.6
% of total	56.2	15.7	27.2	0.90

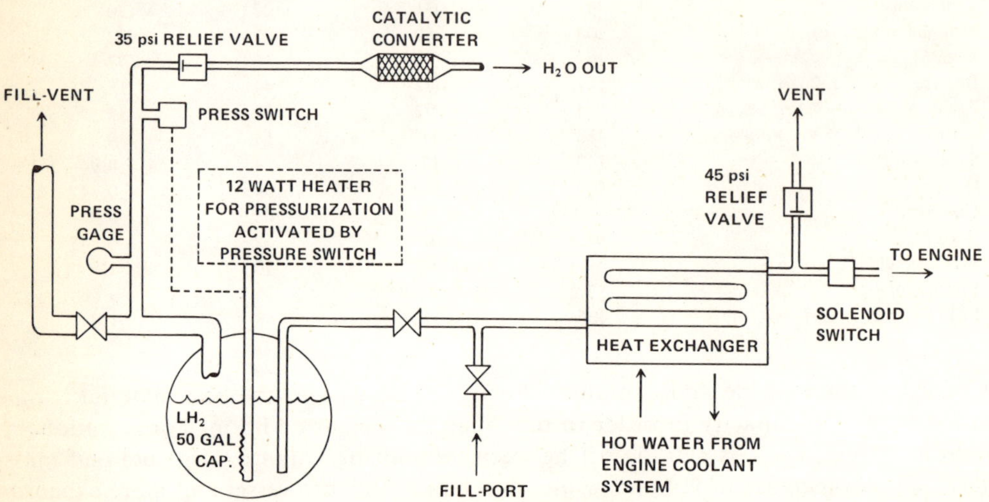

FIGURE 20.   Liquid hydrogen design used at UCLA.[50]

(−424°F). Moreover, the critical temperature of hydrogen is 33 K, and its critical pressure is 14 atm. Clearly, then, the problem of storing liquid hydrogen is one of maintaining extremely low temperatures rather than high pressures. Modern cryogenic technology and equipment make this problem amenable to solution.

A liquid hydrogen storage and delivery system has been developed at the Los Alamos Scientific Laboratories and reported by Stewart and co-workers.[49] A similar system, utilizing an identical tank, was used at UCLA to power a small vehicle furnished for the purpose by the U.S. Postal Service.[50] The key elements are shown schematically in Figure 20.

Hydrogen flows from the tank as a liquid and is vaporized in heat exchange with the engine cooling water system. The heat exchanger is a simple shell-and-tube unit, with the tube bent so that it makes five passes through the fluid in the shell. Hydrogen gas, essentially at atmospheric temperature and pressure, then goes to the engine. A simple on-off solenoid switch in the hydrogen gas line remains off unless energized. This is an added safety feature, no flow being possible until it is desired to operate the engine. A relief valve in the line prevents excessive pressure buildup.

A pressure of approximately 5 psi above atmospheric is adequate to obtain the desired flow rate to the engine. When the pressure in the hydrogen tank drops below this value, a pressure switch automatically activates a heating element in the tank, generating sufficient vapor to increase the required pressure differential. A 12-W heater seems adequate for this purpose.

In time, heat leakage from the atmosphere into the tank results in some boiling of the hydrogen and an increase in pressure. The tank is vented through a check valve

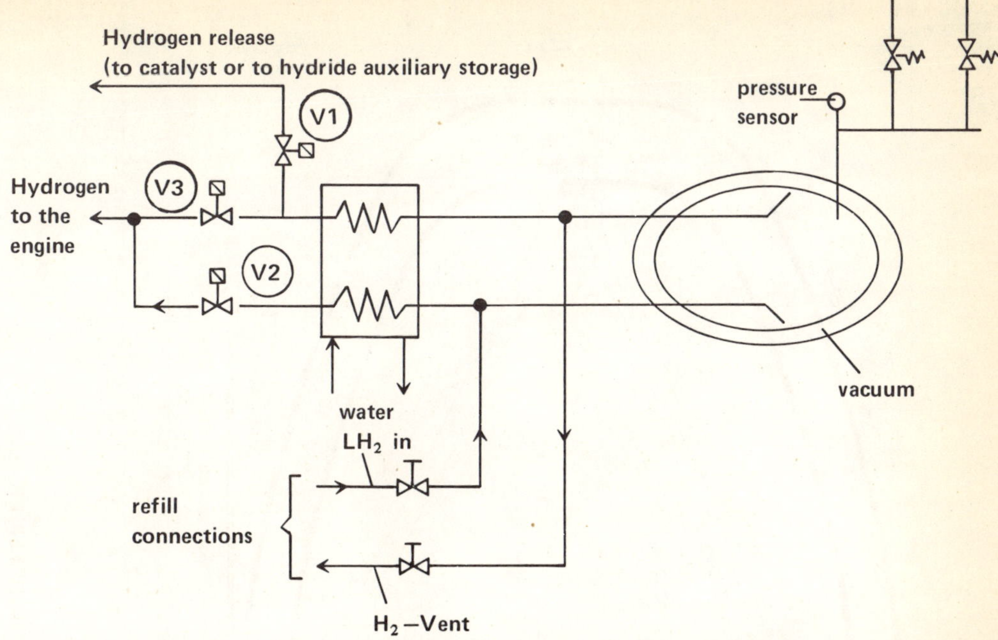

FIGURE 21.   Schematic diagram of LH₂ tank design by the DFVLR.[51]

set at 35 psi to prevent this pressure from becoming excessive. The hydrogen is actually discharged to the atmosphere. A fill vent is also incorporated in the system. This is a larger capacity flow channel and is open during the filling of the tank to permit a more rapid operation.

Further details of the system need not be given here. The essential point to be made is that the system performed well in tests. There was no problem in obtaining adequate flow to the engine, and engine performance was excellent. While in this early version heat leakage into the tank resulted in more hydrogen being lost through venting than desirable, this should not be a problem with successor (improved) tanks. Experience with this system also afforded some "filling station" experience. The tank is filled with the cooperation of the Union Carbide Linde Division plant in Ontario, California. A vacuum jacketed line with bayonet-type fitting is used to connect the fill line of the vehicle to the plant supply. This can be done expeditiously and the filling operation completed quickly if the tank is still cold from a previous fill. If the tank is at atmospheric conditions, purging and chilling is necessary, and the operation may take 15 to 30 min.

The DFVLR - Institute for Energy Conversion in Stuttgart, Germany has designed a cryogenic tank specifically for passenger cars.[51] Their system schematic is shown in Figure 21 for the ellipsoidal tank design shown in Figure 22. A combined liquid and vapor withdrawal system is used to prevent boil-off losses during operation. Hydrogen boil-off during periods when the vehicle is not being operated can be avoided up to about 10 days or more.[51]

The aluminum version of this design is reported to hold 120 ℓ of LH₂ in a volume of 140 ℓ. This gives an energy storage density of 6 kWh/kg with 2.1 kWh/ℓ. The maximum operating pressure is 8 bar (0.8 MPa), with a design operating pressure of 3.5 bar (0.35 MPa). Empty weight of the stainless steel version is 75 kg, while the aluminum version is only 40 kg. A heat leak of 2 W is reported for the aluminum container, resulting in a boil-off of 5.2 ℓ/day of LH₂.

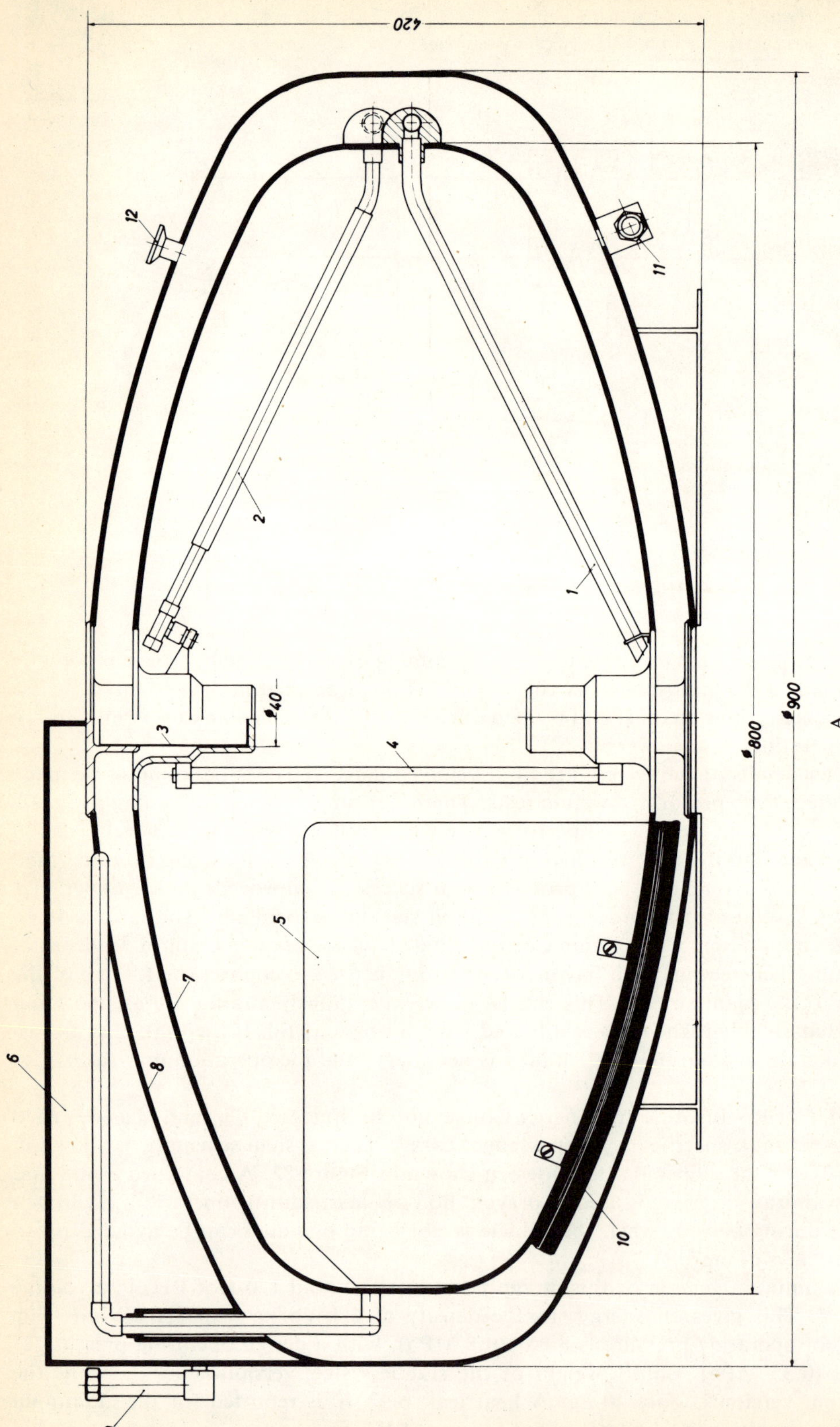

FIGURE 22.   Cross section (A) and photos (B and C) of the DFVLR-LH$_2$ tank.[51]

FIGURE 22B

FIGURE 22C

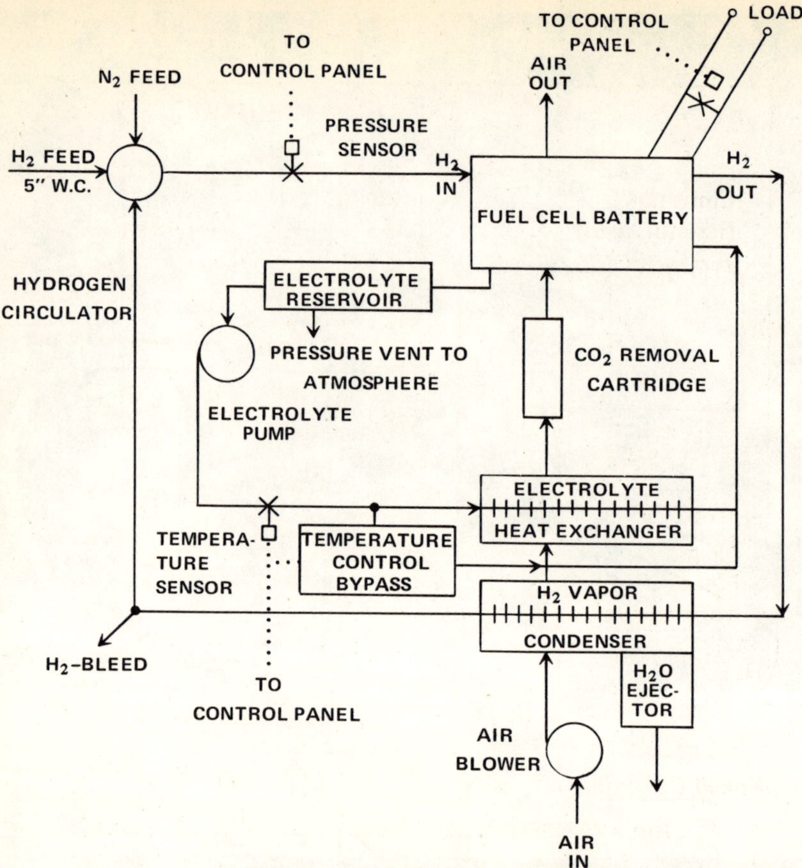

FIGURE 23.   Schematic drawing of hydrogen-air fuel cell automobile.[52]

1.6.2. Prototypes Using Fuel Cells

Fuel cell prototypes have been constructed and tested by Allis-Chalmers between 1959 and 1963, by General Motors in 1964, and by Dr. Karl Kordesch in 1970. An excellent summary of the various types of fuel cells and the technology base is contained in a NASA survey report.[52]

The early efforts at development of a fuel cell vehicle — although highly significant programs — were never able to reach commercialization. Besides the problem of the fuel cell itself, there was also the difficult problem of fuel storage. In a hydrogen system such as the GM® Electrobus,[52,53] the hydrogen was stored as a liquid. Since this cell used potassium hydroxide as the electrolyte, the designers chose to store liquid oxygen as well in order to avoid the problem of contamination by CO_2 in the air.

K. V. Kordesch[52,54,55] was able to use the same cell in an air-breathing design through the use of a CO_2 removal cartridge with an air blower (Figure 23). Hydrogen was stored as a pressurized gas on the top of a 1961 British Austin A-40 chassis, two-door, four-passenger sedan. Another significant feature of the Kordesch design was the use of lead acid batteries to provide acceleration. The electrical schematic is shown in Figure 24.

Kordesch reported few problems with the system during 1 year of operation.[55] This design is of interest today because of its practicality and also because of the advances in hydrogen storage technology — principally, the metal hydrides. The energy effi-

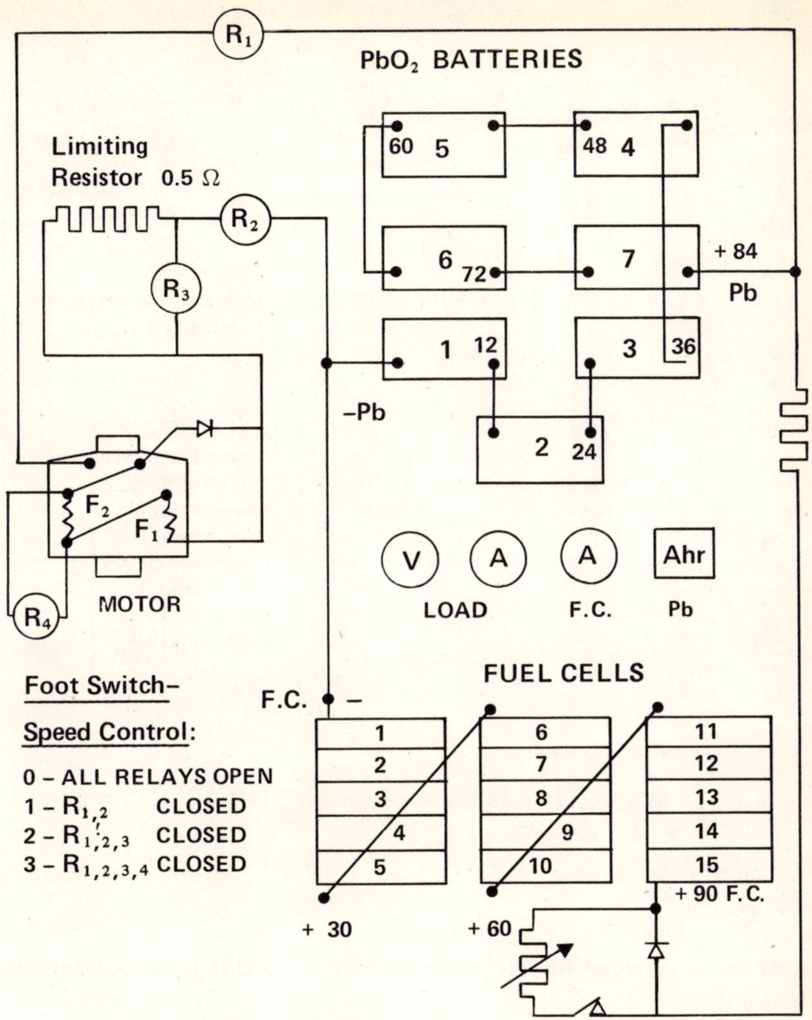

FIGURE 24.   Electrical circuit diagram of fuel cell/battery-powered car.[52]

ciency of the system is high because of the high efficiency of the fuel cell (50 to 60%
with projections to 75%). Operational data on load sharing are given in Table 6. It
may be seen that the battery provides for a large fraction of the high power needs
while the fuel cell is sized for low-speed cruise and battery recharge when the car is
stopped.

The Kordesch design is a hybrid vehicle that could be considered either an electric
battery car or a hydrogen car. It could also be considered a dual fuel vehicle (hydrogen
and electricity) if a means for recharge of the batteries is provided, as in the normal
electric vehicle design. Supply of hydrogen by electrolysis would make the design an
all-electric car.

The technology embodied in the vehicle is quite old, and it is believed that substantial
potential for improvement exists in three areas: fuel cell advancement, battery development, and hydride storage of hydrogen.

A conference sponsored by the Department of Energy was recently held to assess

TABLE 6

Load-Sharing of the Hydrogen-Air Lead System Under Different Operating Conditions (Experimental Data)

Output			Pb battery state		H_2-air	
kW	V	A	of charge	±A	A	Operating conditions
11.0	85	125	1/1	− 80	45	55 mph fourth gear
9.5	80	115	3/4	− 60	55	52 mph reduced field
8.0	75	105	1/2	− 40	65	48 mph battery temperature
6.5	70	95	1/4	− 25	70	42 mph 65°C
20.0	75	260	1/1	−200	60	Steep hill, second gear or start from standing
14.0	65	210	3/4	−140	70	
12.0	60	195	1/2	−120	75	Battery temperature 45°C
8.0	90	90	1/1	− 60	30	45 mph fourth gear
7.0	85	80	3/4	− 40	40	42 mph full field
5.6	80	70	1/2	− 20	50	38 mph battery temperature
4.5	72	60	1/4	0	60	35 mph 65°C
4.0	70	58	1/4	+ 7	65	32 mph fourth gear
3.0	84	35	1/2	+ 8	48	20 mph third gear
2.0	88	22	1/2	+ 13	35	10 mph second gear
0.5	110	5	1/1	+ 5	5	Car stopped
2.0	100	20	3/4	+ 20	20	Charging depends on state of charge and temperature
2.8	95	30	1/2	+ 30	30	
4.2	85	50	1/4	+ 50	50	

Note: 60-A current equals 100 mA/cm² current density.

From Crowe, B. J., NASA SP-5115, NASA, Washington, D.C., 1973.

the prospects for development of a fuel cell vehicle.[56] Several important conclusions were reached:

1. The weight and size of the fuel cell has been reduced to levels reasonable for vehicular applications.
2. The price of fuel cells is projected to drop dramatically in the next 4 years.
3. Basic design improvements combined with mass production should make the fuel cell economically competitive with the internal combustion (IC) engine in the foreseeable future.
4. High energy efficiencies are possible with theoretical projections of 75% and operating cells now at 50%.
5. Fuel cell type depends on fuel availability with the phosphoric acid electrolyte preferred if reformed liquids are used. The KOH cell or solid polymer electrolyte (SPE) aerospace type cells are appropriate for vehicles supplied with hydrogen.
6. Fuel cell and related technologies have advanced to the point that they should be considered for vehicular applications.
7. A hybrid fuel cell/battery vehicle could meet the performance standards of a conventional IC engine while providing far more efficiency with less pollution.

1.6.3. Prototypes Using Metal Hydrides

Hydrogen vehicles that store hydrogen as a metal hydride have been constructed by the Billings Energy Corporation of Provo, Utah and by Daimler Benz of Stuttgart,

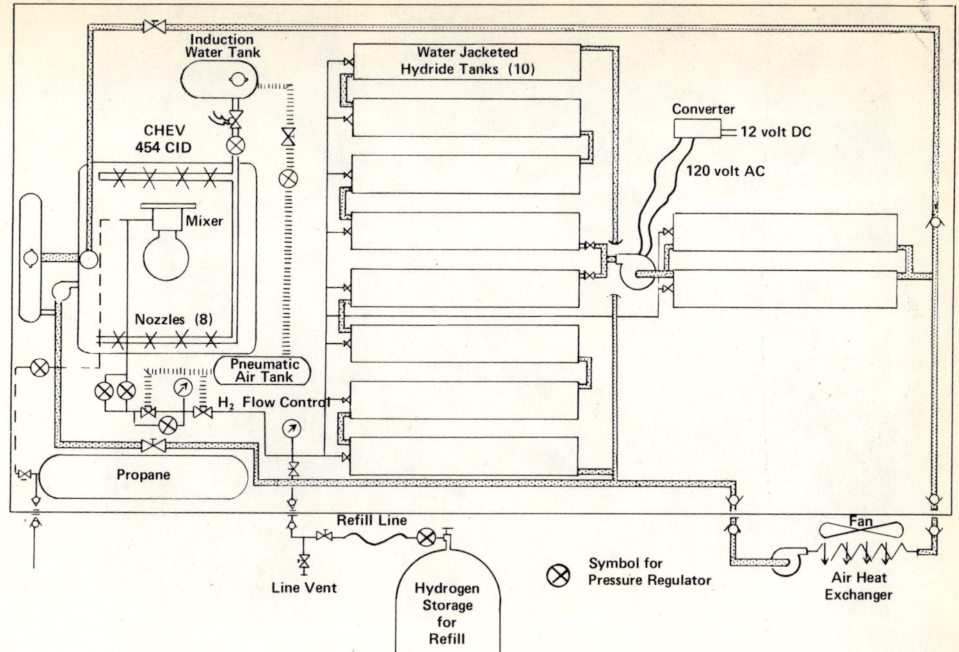

FIGURE 25.   System design schematic for the Riverside hydrogen bus.[45]

Germany. Both companies have test operated several vehicles. The ones included herein were selected as typifying design options.

The first bus produced by the Billings group stored iron-titanium hydride in 3-in. stainless steel tubes that were externally heated by engine exhaust gas.[57,58] Daimler Benz has adopted a similar design and used a combination of magnesium-based hydride and titanium-based hydride.[59]

Later prototypes produced by Billings use engine coolant to heat a titanium-based hydride in an aluminum pressure vessel. The vessel is heated externally from a water jacket surrounding the vessel. A hydride alloy composition as recommended by the Brookhaven National Laboratory has been adopted, $Ti_{51}Fe_{44}Mn_5$. The manganese replaces iron in the usual equal atomic composition. One extra titanium atom replaces iron in this formulation for ease of manufacture. This is close to the formulation known as BNL-191 hydride which has a composition of $Ti_{50}Fe_{45}Mn_5$.

A system typical of the engine coolant design used by Billings is that of the Riverside hydrogen bus.[45] A schematic of the Riverside hydrogen bus is shown in Figure 25. There are six flow circuits in the design:

1.   Hydrogen delivery
2.   Heat exchanger fluid
3.   Induction water
4.   Pneumatic air
5.   Intake air
6.   Propane

Hydrogen is stored in ten hydride vessels positioned beneath the vehicle in the orientation shown in Figure 26. Each tank has a separate valve that may be removed from the system without contamination of the hydride. After manifolding together,

FIGURE 26.   Hydride tank mount and water jacket on valve end of pressure vessel.[45]

the flow passes through a main ball valve that is pneumatically actuated. The valve is open when three conditions exist simultaneously: ignition key on, hydrogen flow switch on, and the existence of manifold vacuum. A regulator then reduces the pressure from the recharge value of 750 psi to 200 psi (5.2 to 1.38 MPa). Heat transfer and hydrogen withdrawal from the hydride are enhanced by having a low withdrawal pressure. Therefore, a regulator bypass circuit has been incorporated that actuates when the tank pressure falls below 200 psi (1.38 MPa). The bypass is another full-flow ball valve with pneumatic actuator. Two low-pressure regulators in parallel are used to adjust the hydrogen pressure to feed the mixer.

The hydride is heated by circulating engine coolant through the water jacket on the tanks. An additional circulation pump was installed to assist the water pump on the engine. This pump is an AC unit supplied with power by a 12-V DC to 120-V AC converter. A parallel/series flow arrangement was adopted in order to have a low pressure drop in the circuit and to simplify the plumbing.

Water that is to be sprayed into the intake is stored in a pressurized water tank. Pneumatic air is used to force the water through the spray nozzles located in the intake manifold. A solenoid valve and regulator is used to activate the spray in accordance with demand as indicated by manifold vacuum. Pneumatic air is provided on this bus to operate the door. Therefore, it was available for use to pressurize the water induction tank and actuate the hydrogen ball valves. Intake air flows through an air filter mounted in front of the radiator and ducted to the mixer. The use of cool air rather than engine compartment air increases the volumetric efficiency.

Propane is stored in a tank beneath the floor just behind the left front wheel well. The propane regulator shuts off tightly until vacuum is applied through a line not shown in Figure 25. It is possible to convert to propane by changing the position of one switch. This can be done while driving; however, only light-load operation is recommended. For normal and heavy operation, it is necessary to replace the welded set

FIGURE 27.   Hydride tank mount and water jacket on base end of pressure vessel.[45]

of distributor advance weights with the standard set of weights. Otherwise, the spark timing is not advanced with increasing rpm.

Refueling the system involves service of three of the above flow circuits. Hydrogen is supplied through a quick connect from storage. The heat exchanger fluid circuit is tapped at two locations with quick connects. An external pump is used to circulate the fluid through an air-blown heat exchanger. When first delivered to Riverside, a water-cooled heat exchanger was used. A change in the design was made because of the drought in California.

It should be noted that the heat created during recharge could be reclaimed for some useful purpose if a fleet of such vehicles were in operation. The amount of heat generated during a single refueling of a bus with 2000 lb (907 kg) of TiFe hydride will be approximately 100,000 to 200,000 Btu 0.1 to 0.2 GJ), depending on the degree of discharge before refueling. The temperature of this heat will depend on the circulation flow rate selected (the same as for a solar collector). For a recharge pressure of 750 psi (5.2 MPa), the hydride equilibrium temperature is about 90°C. For purposes of estimation, it is reasonable to assume that 100,000 Btu (0.1 GJ) may be obtained at a temperature of 65°C each time the bus is refueled. More heat would be available at a lower temperature.

For two refills per day, each hydrogen bus would provide a heat source equivalent to 400 ft² (11 m²) of flat plate solar collector assuming 500 Btu/ft² (5.7 MJ/m²) average daily output. If used for home heating, one bus would provide enough heat for a 1400-ft² home (130 m²) when the outside temperature is 40°F (22°C) colder than the inside.

The remaining flow circuit that must be serviced during the refueling operation is the water induction reservoir. This must be level checked and refilled as necessary. Several photographs are included to help describe the system. Figures 26 and 27 show the construction of the hydride storage vessel. The water jacket and support structure are combined. The ten vessels are supported at the ends and suspended beneath the floor. In this design, the precertified aluminum tank is encapsulated without welding to the tank itself. The aluminum pressure vessel is commercially supplied by Luxfer USA Limited of Riverside, California. The selected vessel is the N-150 of length 48.2 in. (122.2 cm), 8-in. OD (20.3 cm), and 0.362-in. wall thickness (0.92 cm). The internal

FIGURE 28.   Ten water-jacketed and insulated hydride tanks ready for installation in the Riverside hydrogen bus.[45]

volume is 1.04 ft³ (29.45 ℓ). Each tank weighs 49.8 lb (22.6 kg). The service pressure is 2015 psi (13.9 MPa).

The physical size of the ten completed tanks is shown in Figure 28. Careful consideration was required in tank selection and jacket design in order to match the available space along the frame. It was necessary to relocate some hardware.

The Riverside hydrogen bus is the first vehicle to be operated in regular passenger service by a transit authority (Figure 29). The program, which is funded by the state of California through CALTRANS, is intended to obtain operating experience with a hydrogen vehicle. This test program is providing valuable information relative to reliability, maintainability, and service operation. Feedback from the transit operator to the conversion contractor as to the performance of the system has proved to be highly valuable. The suggestions have been heeded to modify the design and substantially improve the reliability.

Billings has also demonstrated a dual fuel hydrogen/gasoline system as part of the Hydrogen Homestead.[60] The family car (Figure 30) is a 1977 Cadillac Seville. It is well known that the range of the car of this size is limited by the weight of the installed hydride tank. Research into improved hydride materials continues. In the interim, the practicality of a dual fuel system is demonstrated by the hydrogen Cadillac. Since a large fraction of road miles is accrued in short trips, i.e., commuting to work and shopping errands, a small hydrogen storage system that is switchable to gasoline service is satisfactory at present. Switching between hydrogen and gasoline may take place while driving this vehicle. Garage recharge of the hydride tank from a Billings electrolyzer permits predominant hydrogen utilization in the vehicle for a typical driving

FIGURE 29.   Ribbon-breaking ceremony when hydrogen bus was first delivered to Riverside, California in August 1977.[45]

cycle. Longer trips are taken using gasoline. The hydrogen Cadillac dual fuel system is shown in Figure 31.

A Jacobsen garden tractor (Figure 32) is powered exclusively by hydrogen.[60] This vehicle is illustrative of the practicality of hydrogen fuel for work vehicles operated in proximity to a source of hydrogen. The system design of the tractor is simplified as much as possible (Figure 33). The prototype is interesting for this reason and also because the engine is air cooled. The low-temperature engine coolant system shown in other Billings prototypes was adapted in this case by placing a heat exchanger inside the exhaust muffler. In service, this design provides for rapid heating of the water jacket surrounding the hydride. Cold startup with hydride approaching complete discharge has been accomplished many times. An exhaust heat exchanger such as this could be used to supplement the heat obtained from the engine coolant.

A summary of recent projects conducted by Billings was made by Ruckman et al.[61]

Daimler Benz has designed a system that performs four functions simultaneously:[59]

1.   Fuel storage and supply
2.   Air conditioning
3.   Water condensation
4.   Waste heat storage

They also are test operating a hydrogen/gasoline dual fuel automobile[62] (Mercedes Benz PKW, type 280E).

Several hydride tank designs were tested in the van shown in Figure 34. A conventional 2.3-ℓ, four-cylinder engine was converted to hydrogen without any major modifications (power output = 44 kW, 4800 rpm, CR = 9, 25% exhaust gas recirculation).[59]

The early design used by Daimler Benz uses a titanium-based alloy heated by engine coolant (Figure 35). The 65-ℓ vessel holds 200 kg of TiFe hydride. It was located behind the front seat beneath a bench.

FIGURE 30.   Billings hydrogen/gasoline Cadillac.[60]

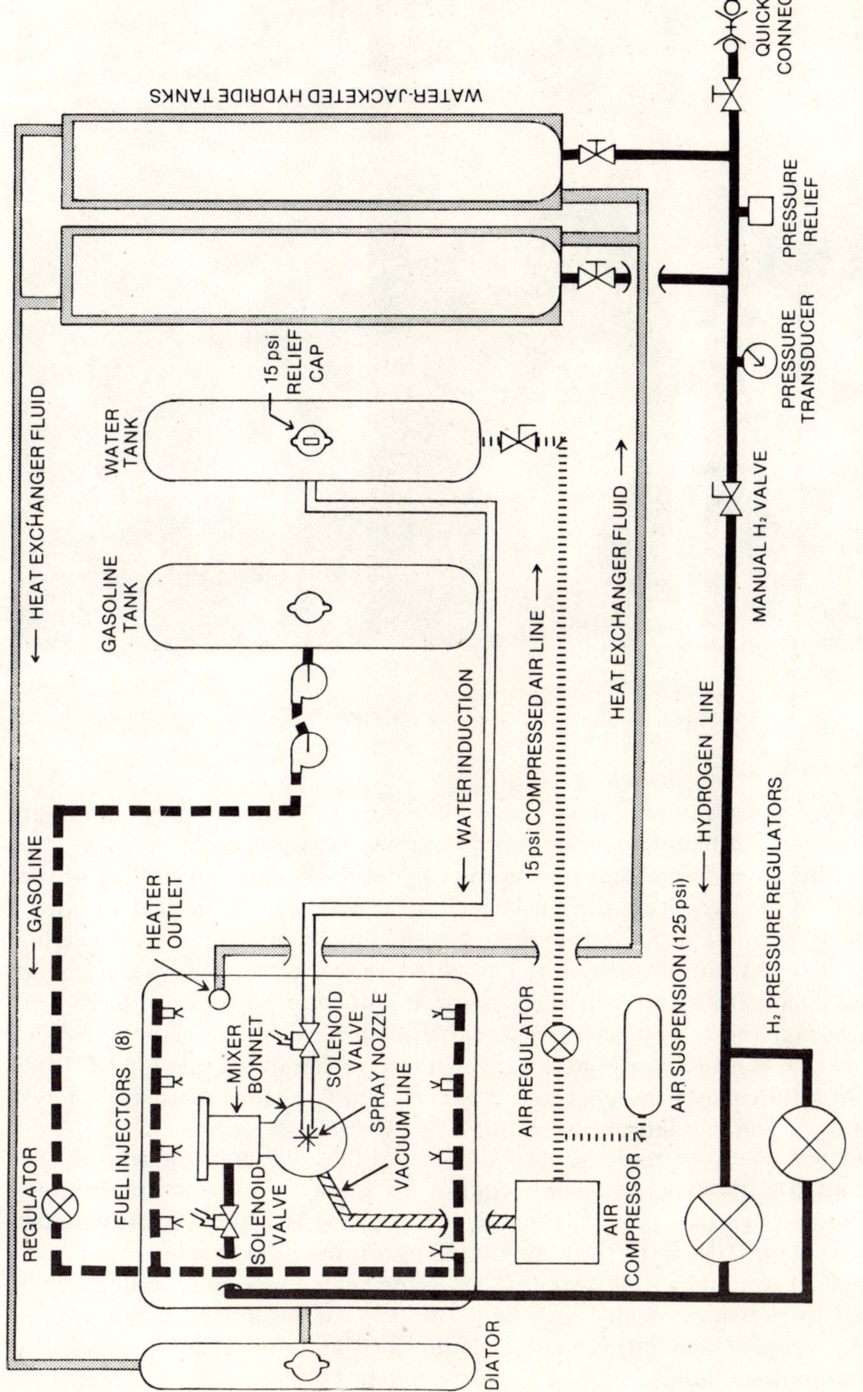

FIGURE 31. Hydrogen/gasoline dual-fuel vehicle.[60] Billings Energy Corporation, Provo, Utah. Patent pending.

FIGURE 32.   Hydrogen lawn tractor.[60]

Results of a typical test run are shown plotted in Figure 36 with tabulated data given in Table 7. The working temperature of this design was 80°C. Prior to recharge, the hydride was cooled down to ambient temperature by means of external cooling water. Then up to 75% of the overall H_2 storage capacity was absorbed in the first 10 min.[59] A strategy was devised to further reduce the recharge time by interrupting the heating cycle while the van was still in operation. The objective was to reach the refuel point with hydride at a temperature of −20°C. A suggested mode of operation is to control the flow of heat exchanger fluid such that the hydride is always kept between −20°C and +20°C. No tank insulation is necessary at these moderate temperatures. Tank pressure should ideally stay between 1 and 2 bars (0.1 to 0.2 MPa). Figure 37 shows test results with a system that interrupts heating of the hydride.

A further advantage of the mode of operation is that the cold tank can be used as a heat sink for air conditioning. Such use requires no energy input from the engine. Figure 38 shows a design that takes advantage of this property of FeTi hydride. Heat input to the hydride (SP1) comes from air that is blown over the tubular containers. At the same time, a second hydride tank (SP2) is being heated by engine coolant while flow is shut off by pressure switch D2. When tank SP1 can no longer sustain a flow sufficient to meet engine demand, vessel SP1 is shut off, and the valve for vessel SP2 is opened. Pressure then builds in SP1 to approximately 1 bar, and the cycle repeats. In this way, both tanks remain cold for air conditioning purposes. When tested, the van had a range of approximately 200 km for an FeTi hydride weight of 300 kg.[59] The

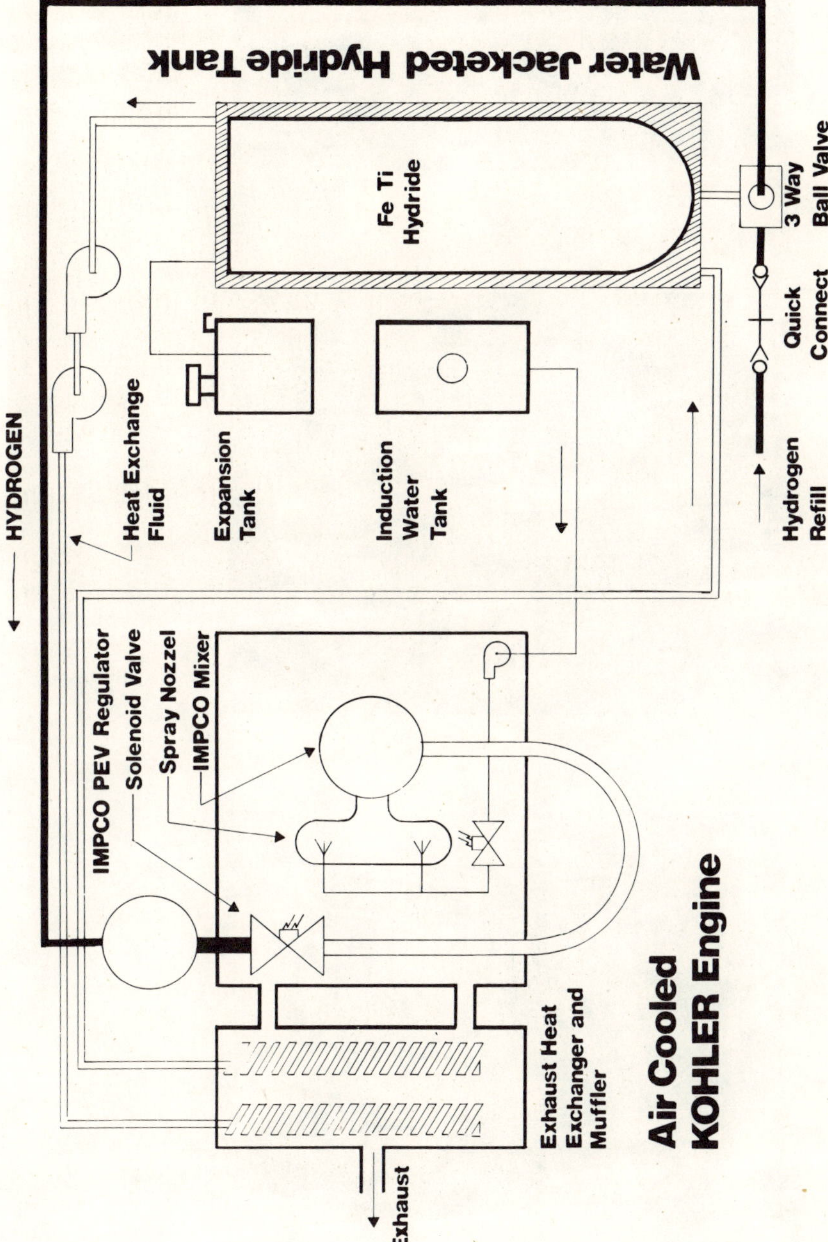

FIGURE 33. Schematic drawing of hydrogen conversion of the Jacobsen Tractor and Kohler Engine.[60] Billings Energy Corporation, Provo, Utah. Patent pending.

A

B

C

FIGURE 34.   Hydrogen test vehicles used by Daimler Benz (Mercedes Benz) : (A) van used for several experiments, (B) gasoline/hydrogen dual-fuel car, (C) hydrogen city-bus featuring a hydride air conditioner.[59,62]

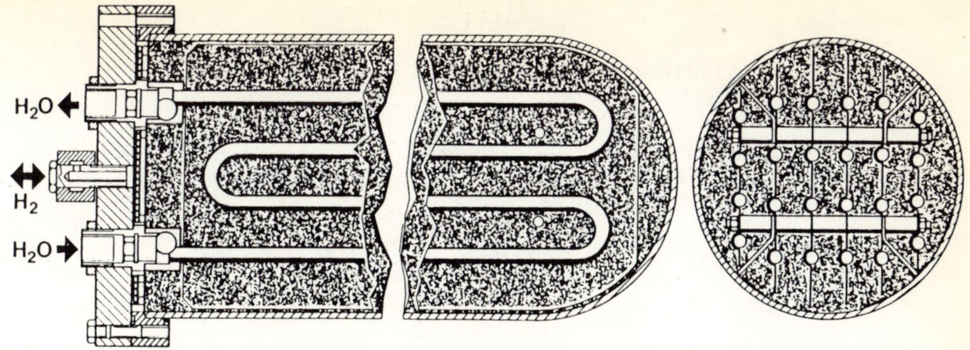

FIGURE 35.   Hydride container tested in Mercedes Benz van.[59]

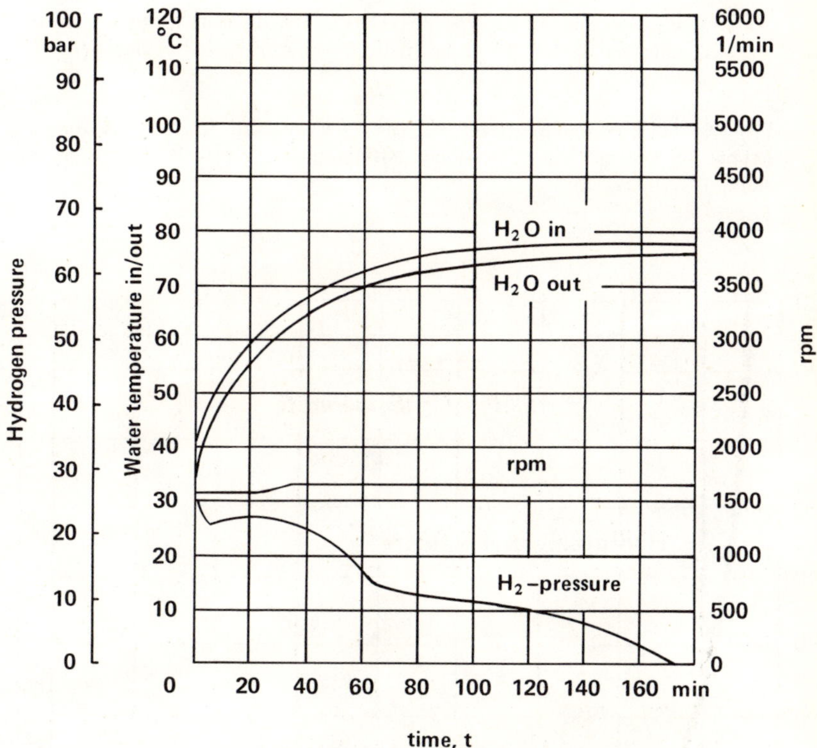

FIGURE 36.   Performance of TiFe hydride vessel shown in Figure 35.[59]

heat exchanger design was such that all of the heat could be taken from the engine coolant when air conditioning was not needed.

Titanium-based hydride has a low storage capacity of about 1.6% by weight (1.95% maximum). Magnesium-based alloys have much greater capacity but require dissociation temperatures between 200 and 400°C to deliver hydrogen to the engine at pressures greater than atmospheric. The heat required by the hydride reaction is about twice that of TiFe. A design was therefore adopted that used both hydrides as shown in Figure 39. Exhaust gas passes through a matrix of tubular containers containing Mg$_2$Ni hydride and then through an array of TiFe containing tubes which can be designed for

TABLE 7

Test Data of Second Experimental Hydride Vehicle

Test	Hydrogen consumption (m³)	Distance (km)			Driving mode		
		a	b	c	km/h	1/min	Gear
1	32	88	132	165	60	2200	4
2	40	86	105	130	70	2600	4
3	45	81	87	110	80	3000	4
4	48	76	76	95	60	3500	3
5	48	79	79	100	30	1600	3
6	48	66	66	83	30	2700	2
7	48	69	69	86	70—104	M a x 4000	4
8	48	130	130	165	60	2200	4

Note: a = distance recorded; b = distance calculated for 48 m³ hydrogen; c = distance calculated for 48 m³ hydrogen and optimized engine (with consumption reduced by 25%).

From Buchner, H. and Saufferer, H., NATO/CCMS 4th Int. Conf. Automotive Propulsion Systems, Vol. 2, Session 6, Arlington, Va., April, 1977. With permission.

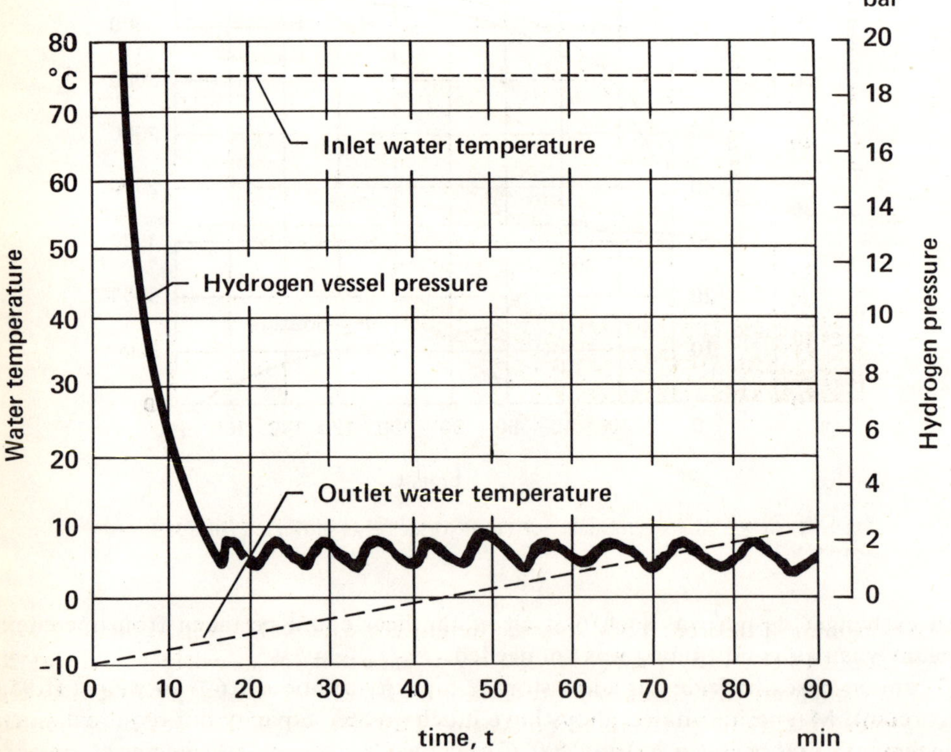

FIGURE 37.   Test data of TiFe hydride vessel (Figure 35) indicating cylical withdrawal of hydrogen to maintain a low hydride temperature.[59]

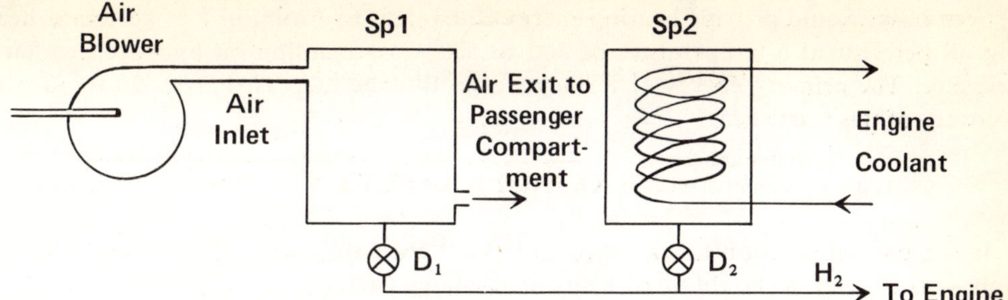

FIGURE 38.   Hydride air conditioner unit developed by Daimler Benz.[59]

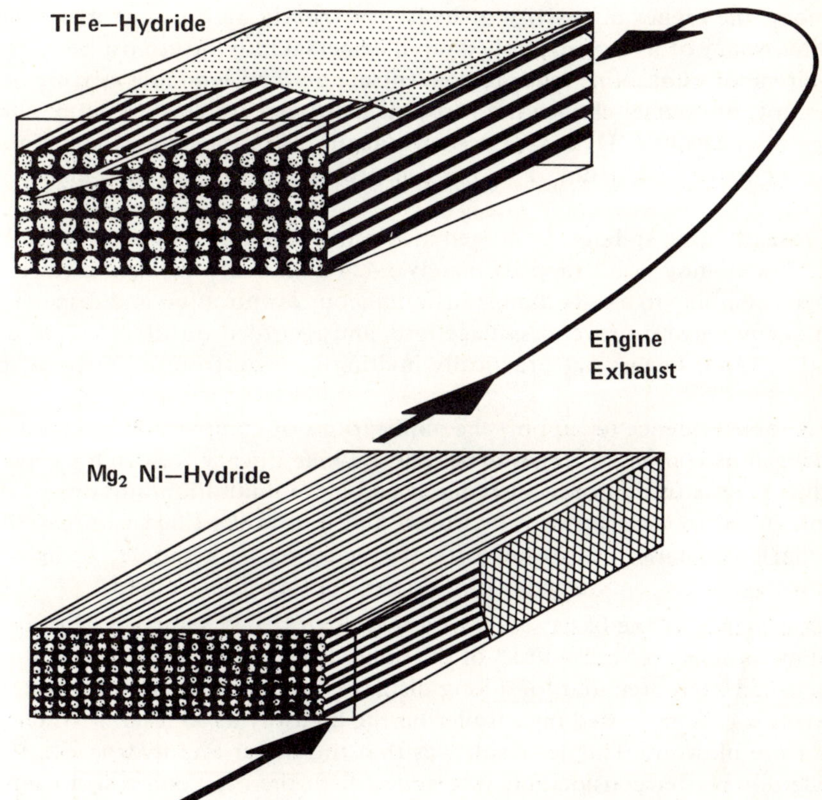

FIGURE 39.   Dual hydride system using exhaust heat (system used in Mercedes Benz city bus).[63]

air conditioning if desired. Total hydride weight consists of 1/3 Mg_2Ni and 2/3 TiFe. When placed in the van, a range of 400 km is expected for a hydride weight of 200 kg. Test data were scheduled for presentation at the 2nd World Hydrogen Energy Conference in Zurich, Switzerland on August 21 to 24, 1978. The design as outlined earlier[59] will absorb nearly all of the energy in the exhaust, thus making possible a nonheat-emitting vehicle. Such a system is advantageous for mining applications. It also means that the waste heat of combustion may be reclaimed at the refueling station as low-grade heat. Estimates made by Daimler Benz[59] indicate that a fleet of 300 hydrogen buses would provide heating energy equivalent to 2 million ℓ of kerosene heating oil per year at a temperature of 300 to 500°C, depending on hydrogen recharge pressure. The primary energy efficiency would thus be increased from 20 to 50% by recovery of this waste heat.

1.7. VEHICLE SAFETY

It is a natural reaction to be concerned about the safety of hydrogen-fueled operation — as, indeed, it should be with any high-energy fuel.

It has been noted that hydrogen ignites readily and has a wide range of flammability limits. Thus, in the event of a leak, it might be feared that hydrogen is much more apt to ignite than, say, gasoline. Note from Table 1, however, that the lower limit of the hydrogen-air flammability range is above the corresponding point for gasoline. Furthermore, the higher diffusivity of hydrogen (nearly an order of magnitude) and the great buoyancy of hydrogen imply a rapid dispersal into the atmosphere. The heavier-than-air vapor of gasoline tends to concentrate until an ignitable mixture develops. This does not, of course, change the fact that hydrogen systems are more vulnerable to leakage than gasoline. Hydrogen systems (again, as in the case of any high-energy fuel) must be made leak proof. This has not proved to be a problem in the author's experience.

In the case of liquid spillage, hydrogen may prove less hazardous than gasoline. The extremely low boiling point (approximately −424°F or −253°C) and great buoyancy of hydrogen combine to assure almost instantaneous evaporation and dispersion. This has been documented by several spillage tests and recorded on film. Liquid gasoline, on the other hand, lingers on, practically inviting ignition from a variety of possible sources.

There is some evidence to support the implications of comparable or superior safety with hydrogen as compared to gasoline. The Billings Energy Research Corporation[58] has conducted tests firing armor-piercing incendiary ammunition into one of their hydride tanks filled to capacity with hydrogen and into tanks filled with gasoline. The comparison is dramatic. The result with the hydrogen is a simple, straightforward burning; with gasoline, a destructive rupture and conflagration results.

Further evidence of the basic safety of a hydrogen fueled vehicle is afforded by the results of an unplanned "crash test" of the liquid hydrogen vehicle described earlier.[50] The vehicle had been prepared for a long-distance rally, and its tank was full of liquid hydrogen. It was being towed on a trailer on the highway at 55 mph when the tow car suffered a tire blowout. The net result was that the trailer and hydrogen car came to rest upside down after considerable interaction with the road and a sign post or two. There was no fire and no unintentional leakage of hydrogen. In fact, the car was separated from the trailer (the latter being damaged beyond repair), righted, and driven back toward UCLA as far as possible before running out of hydrogen.

These arguments and statements must not, of course, be taken to minimize the need for proper handling of hydrogen — or of any other fuel. It is argued, however, that with proper handling, hydrogen is no more inherently hazardous than gasoline. It may, indeed, be less so.

1.8. RAILROADS

Hydrogen has been considered for use by the railroads. Under sponsorship of the Department of Energy,[64] contacts were made with railroad organizations and individuals in government, industry, and research organizations. The feasibility and desira-

TABLE 8

Comparisons of Characteristics of Storage Alternatives (3200-Gal Diesel Energy Equivalent)

Storage method	Weight (gross) (lb)	Weight ratio to liquid H_2 storage	Cost ratio to liquid H_2 storage	Proven in rail use	Adequate supply of equipment to support program
Iron-titanium hydride car	920,000	11.6	10.4	No	No
Cryogenic storage car	79,000	1.0	1.0	Yes	Yes
3500-psig gas storage car	1,340,000	16.9	5.41	No	No

From Foster, R. W. and Escher, W. J. D., DOE Report CONS/4707-1, U.S. Department of Energy, Washington, D.C., 1976. With permission.

bility of converting conventional diesel-electric locomotives to hydrogen operation was evaluated.

Three storage options were evaluated: iron-titanium, cryogenic liquid, and 3500-psig compressed gas. Only liquid hydrogen was considered feasible at this time (Table 8).

A typical locomotive (Figure 40) uses a turbocharged diesel engine to generate electrical power for traction motors. Engine assemblies come in eight-cylinder configurations and are incremented in two-cylinder groups up to 20 cylinders. Eight-cylinder engines are naturally aspirated while the 12 cylinders and above are turbocharged. Power output ranges from 1000 to 3000 hp.

A breakdown of the major components of the power system is given in the study.[64] Particular attention was given to the general prospect for engine conversion. It was concluded that conversion of an existing engine to hydrogen, using liquid hydrogen and high-pressure injection would provide:

1. Increased power levels of 15 to 20% due to basic cycle changes with hydrogen, and another 15 to 20%, were series turbocharger air-charge cooling incorporated (providing a total of 32 to 44% gain in potential output).*
2. Increased overall thermal efficiency up to a maximum of the order of 5%, if the engine utilizes exhaust heat recovery and/or is further leaned out. (Otherwise, a 1 to 2% reduction in overall efficiency is predicted due to basic thermodynamic effects.)
3. Virtual elimination of smoke, odors, carbon monoxide, and unburned hydrocarbons. Oxides of nitrogen would remain, but several avenues for reducing levels have been identified.

The principal technical unknowns at this time were posed as questions in the report that will be answered in the course of a proposed project:

1. Can reliable compression ignition be achieved? If not, what alternative ignition system approach is most suitable?
2. What hydrogen admission technique is preferable: low- or high-pressure, low- or high-temperature, early or late injection, etc.?

* It is recognized that commensurate peak cylinder pressure increase might then occur. An obvious approach would be to then further lean out the engine to improve specific fuel consumption and reduce NO_x.

FIGURE 40.   E-series locomotive by General Motors Electromotive Division. [64]

3.   What is the most advantageous and practical approach for utilizing the refrigeration capability of liquid hydrogen: air-charge cooling, exhaust gas recuperation, engine coolant heat exchange, etc.?
4.   For selected configurations, what will be the performance (thermal efficiency, power) over the engine rated operating range and at off-design conditions?
5.   What will be the effect on engine and power system reliability and maintenance requirements?
6.   What will be the level of oxides of nitrogen (the only emission of concern), and how can this be reduced if judged too high?
7.   What will be the ramifications to ancillary and auxiliary power system requirements?

Liquid hydrogen tank cars have been in use for many years. A 28,000-gal (106,000-ℓ) tank car used commercially by Linde Division of Union Carbide is shown in Figure 41.

Given the fact that a fuel storage car already exists, the study recommended development of engine conversion methods and demonstration of a hydrogen locomotive.[64] The test locomotive system was envisioned as consisting of a modified E-series locomotive with the rear engine station incorporating a hydrogen fueled, 12-cylinder, turbocharged diesel engine and generator. The locomotive would be followed by a liquid hydrogen tank car (either the Linde 28,000-gal LH_2 railcar or the NASA 34,000-gal LH_2 railcar; the NASA railcar was recommended in the study). Following the tender during the period of development and demonstration would be a modified 85-ft coach car providing crew support and instrumentation capability (Figure 42). Photographs of each of a scale model of each of these three cars are shown in Figure 43.

The development of a hydrogen locomotive would allow the railroads to move away from dependency on petroleum-based fuel. Although there are not now any regulations on railroads, there is concern for the environment, and efforts are being made by the railroads to minimize pollutants, especially smoke.

The study also examined potential costs for liquid hydrogen. It was concluded that LH_2 is significantly more expensive than is diesel oil today, but that the cost of gaseous hydrogen is likely to be comparable to that of both hydrocarbon synthetic fuels and electricity in the future. The incremental cost of liquefaction must be compensated for

FIGURE 41. LH$_2$ railroad tank car (28,000 gal LH$_2$).(Courtesy of Linde Division of Union Carbide.)

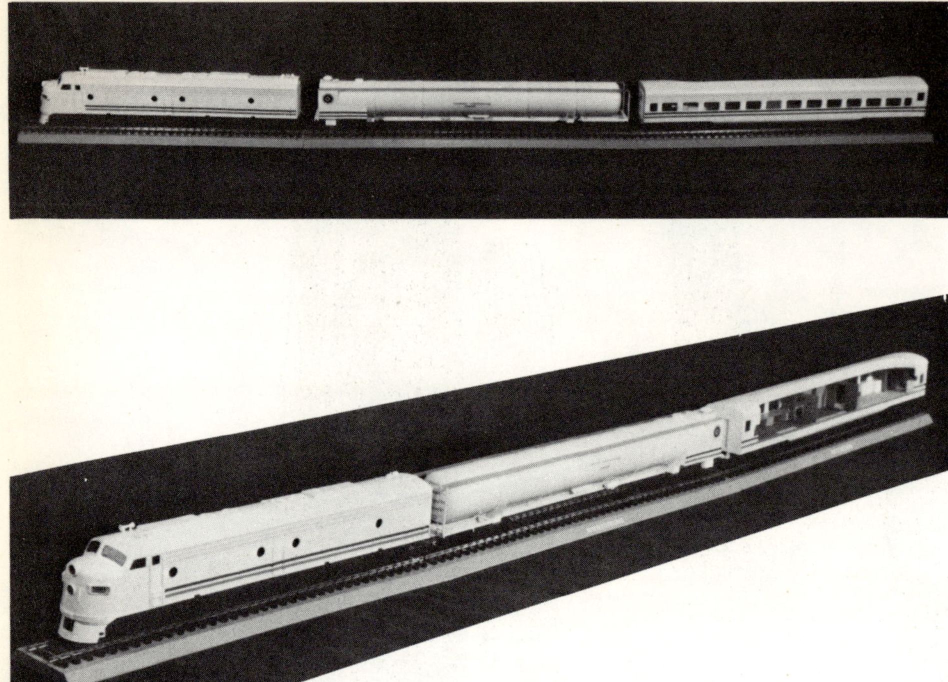

FIGURE 42.   Proposed LH$_2$ train demonstration and development configuration.[64]

by performance, efficiency, and improvements or cost savings in other areas if the system is to be viable.

With regard to technical and operating feasibility, the study concluded that locomotive diesels can be converted successfully to hydrogen. After conversion, the engines are anticipated to have higher power output at a moderate efficiency improvement, longer life, and to be nonpolluting with the exception of NO$_x$, which may be reduced significantly.

The final conclusion of the Department of Energy report was that there exists a third alternative for moving the railroads away from petroleum dependency. Previously, only use of syn fuels and railway electrification were thought to be available. This alternative requires development and demonstration in order to be considered viable by the railroad community.

1.9. COMPARISON WITH ELECTRIC BATTERY VEHICLES

The problems associated with hydrogen vehicles, in particular with those using metal hydrides, are problems similar to those encountered in the design of electric battery vehicles. Energy storage density and volume, recharge time, nonpolluting operation, peak power output, efficiency, range, capital cost, reliability, and service life are many of the parameters that these two systems have in common. Both are nonpolluting alternatives to hydrocarbon fuels and, as such, would be expected to find application in similar product areas. Electrics are now commercially available for use in enclosed environments where emission of pollutants is intolerable. Such applications are forklift trucks, front-end loaders, indoor sweepers and other indoor work vehicles, mining

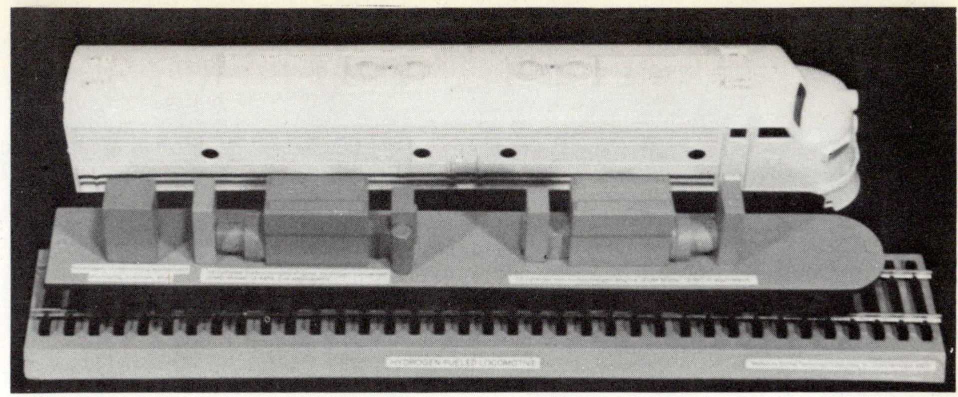

FIGURE 43a. E-series locomotive showing dual-engine locations. Reference 64 proposes to modify the rear engine for hydrogen service.

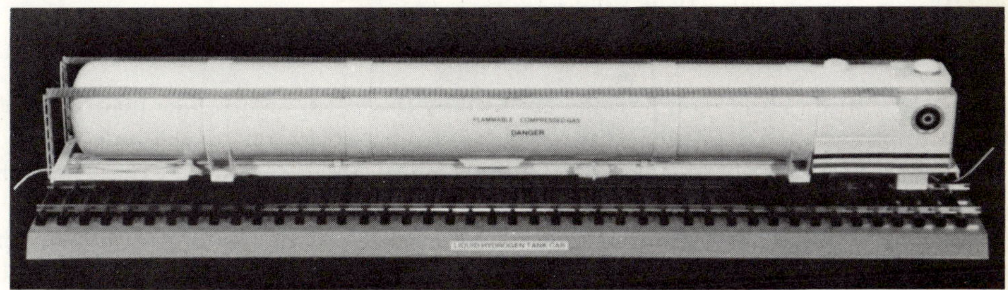

FIGURE 43b. Linde/NASA 34,000-gal LH$_2$ railroad tank car.

FIGURE 43c. Proposed configuration of instrumentation and support car.[64]

equipment, and tunneling equipment. Since the electric battery car has found a limited niche in commuter and shopping applications despite obvious technical inadequacies when compared to the gasoline compact car, it is reasonable to presume that a hydrogen vehicle with similar design difficulties could also be developed along similar lines.

A comparison of the two vehicle types is inevitable. The task, however, is very difficult. The two technologies are both viewed as being immature at the present time. While the battery technology has been in existence many years, the mainstay of the industry — the lead-acid battery — has long been viewed as inadequate storage. There-

TABLE 9

Comparison of Energy Storage in Hydrogen and Electric Vehicles

	Lead-acid battery	FeTi $H_{1.60}$ (typical)	FeTi $H_{1.95}$ (Maximum)	Ratio of hydride to battery
Energy storage density (Wh/kg)	20—30	514 (LHV)	625 (LHV)	17—31
		608 (HHV)	740 (HHV)	20—37
Cost ($/kWh)	45	9.70 (LHV)	8.00 (LHV)	0.22
		8.20 (HHV)	6.80 (HHV)	0.18
Storage efficiency	65%	100%	100%	1.5
Service lifetime	500 cycles	Infinite	Infinite	00
Salvage value (% of first cost)	10%	100%	100%	10
Motor efficiency	75%	30%	30%	0.40
Propulsive energy storage density (Wh/kg)	15—26	154	188	6—12
Comparative range for identical vehicle weight	1	6—10	7—12	
Energy efficiency starting from coal	12%	16%	16%	1.3

Note: LHV = lower heating value; HHV = higher heating value.

fore, battery studies and comparisons are usually based on projections involving various assumptions as to the probability of success for future battery developments.

The hydrogen technology is based essentially on the space program experience with liquid hydrogen. A production and distribution system to supply vehicles with hydrogen does not now exist. Therefore, cost comparisons reflect the uncertainty on this aspect of the problem as well as on the hydrogen vehicle itself. Hydride research has been conducted under limited funding for only the last 10 years (with the exception of some classified work that is generally unavailable). Vehicle programs using both hydride storage and liquid storage have been carried forward by private individuals, project teams of university students, and, most recently, by small organizations within a company. The science of hydrogen use for surface transportation is most assuredly in its infancy. Lacking the existence of a solid basis for the hydrogen vehicle, any comparison with the battery electric vehicle is, at best, difficult.

A cursory comparison of just the energy storage system points to an opportunity for hydrogen vehicle development and commercial application in areas where the electrics are now making inroads. Table 9 shows a comparison of the most widely used battery (lead-acid) with the most widely used hydride, FeTi (which is also one of the heaviest but has desirable thermodynamic properties). On the basis of this primative comparison, the hydride has the advantage by a significant margin in energy storage density, first cost, salvage value, service life, storage efficiency, storage density of energy actually providing vehicle propulsion, vehicle range, and energy efficiency starting from coal as the energy feedstock.

Obviously, a valid comparison is not this simple. Having entered several caveats concerning the difficulty of the task, we present below the results of a study sponsored by the Department of Energy and managed by Mr. James J. Donnelly, Jr. of the Aerospace Corporation.[31]

Under sponsorship by the Alternative Fuels Branch of the Division of Transportation Energy Conservation, Department of Energy, the Aerospace Corporation conducted a cost-benefit study of the relative attractiveness of hydrogen and electricity in automotive propulsion systems for the period of 1985 to 2000. The generation of hydrogen and electricity from various common sources has been studied together with the use, respectively, as a fuel for hydrogen vehicles with IC engines or in battery propulsion systems. The comparison includes consideration of energy feedstocks, production processes, conversion efficiencies, costs of capital equipment and product, transmission, storage, and distribution. Both the energy impact and the cost impact were examined.

An assessment was made of both and hydrogen and battery technologies as they now exist. Based on this assessment and the expressed opinion of researchers in the field, a projection was made for three time frames; 1985, 1990, and 2000.

For the hydrogen vehicle, detailed attention was given to the novel problems of spark ignition engine operation and storage of the fuel in cryogenic liquid and metal hydride forms. Engine-operating problems of flashback, power loss, and NO_x emissions were addressed, and the effectiveness of corrective measures was assessed. Projections were made for engine efficiency, horsepower per unit displacement, and engine weight and cost. Descriptions were provided for near-term iron-titanium hydride and representative advanced magnesium alloy hydride storage systems, including hydride characteristics, system design and integration, tank design, refueling problems, and system cost. Cryogenic storage and feed system topics included system design, tank technology, weight, cost, safety, and refueling. The utility of liquid hydrogen and certain metal hydrides as heat sinks for simple automotive air conditioning systems was noted.

Electric vehicles were similarly assessed, starting with consideration of near-term and representative midterm and advanced batteries. The treatment includes battery descriptions and development goals (specific energy, specific power, efficiency, cycle life, installed cost); technology of drive motors, controllers and other components; auxiliary power requirements; driving cycles; projections of aerodynamic and rolling resistance characteristics; regenerative braking; material, manufacturing, maintenance, safety, and environmental factors; and cost of ownership. Vehicle driveability criteria were established, and performance parameter values were selected for the three selected points in time for the subsequent comparison with the hydrogen-powered vehicle.

The final major task was the comparison of hydrogen and battery-powered passenger cars. Overall energy efficiency comparisons were first made considering alternative fossil and nonfossil sources, conversion to gaseous or liquid hydrogen and electricity, transmission, storage and distribution by various modes, and charging into vehicle hydrogen tanks or batteries. Analyses of weight, size, and cost were then made using comparable parametric performance criteria for hydrogen and electric automobiles. Even in their more advanced designs, these vehicles were found to be decidedly more range and performance constrained than conventionally fueled cars. Consequently, design range, on a selected driving cycle, was varied to accommodate trade-offs between vehicle usefulness, weight, performance, and cost of ownership. Sensitivity analyses included the effects of auxiliary power loads, regenerative braking recovery factors, and aerodynamic drag characteristics. Vehicle energy consumption characteristics were then combined with the previous energy source analyses to yield the overall efficiency from source through propulsion mission.

Detailed lists of the various parameters selected for use in this study are available in the report.[31] The SAE J227 schedule D driving cycle was chosen for the parametric analysis. There is, of course, a wide variance of opinion as to the appropriateness of the values selected. In particular, the projections for development are subject to large uncertainty. The study is, however, nonbiased and reasonably thorough.

Figure 44 gives an estimate of vehicle weight, length, and usefulness (statistical percentage of the number of days in the year that range is adequate) vs. the vehicle range. Limited range is necessary for both battery and hydride vehicles. The LH_2 vehicle is able to attain the range characteristics of the typical automobile.

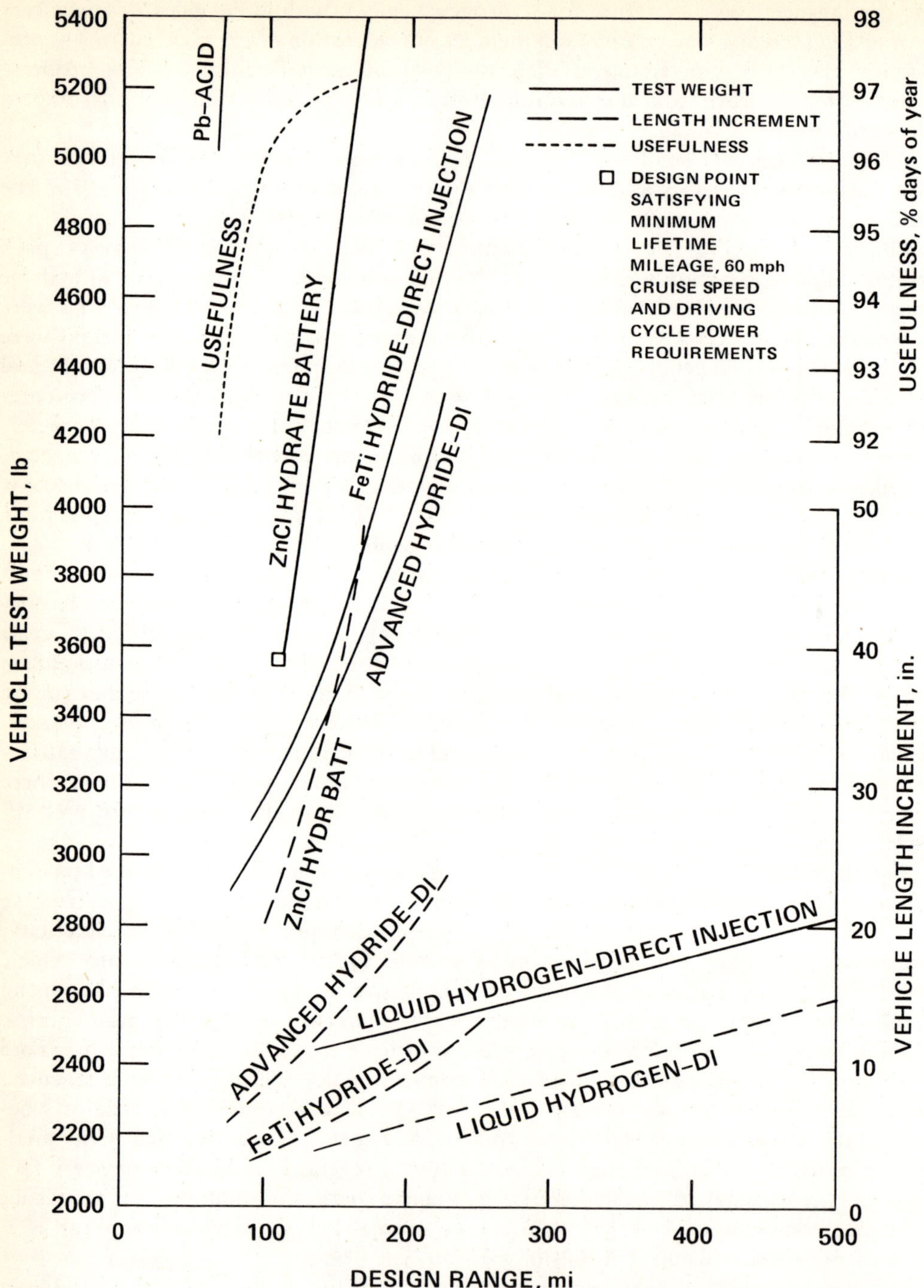

FIGURE 44.   Comparison of hydrogen and electric battery vehicles for a given design range (1990 model year assumed).[31]

Tables 10 to 15 and Figures 44 to 46 summarize the results of the study. Table 15 is the final summary of the report for all vehicles and conditions investigated. Key characteristics of all minimum design range vehicles are presented along with the implications of increasing the range to 200 mi. Following are the principal features projected for each vehicle type considered in the Aerospace study.[31]

TABLE 10

Summary of Electric Vehicle Design and Performance Characteristics

Model year: battery

	1985: Lead-acid	1985: Nickel-zinc	1990: Zinc chlorine hydrate	2000: Lithium aluminum iron sulfide
No. batteries required for 100,000 mi	2	2	1	1
Battery weight (kg/lb)	887/1956	500/1102	358/789	175/386
Vehicle resistance characteristics, C_R/C_DA [(lb/lb)/ft^2]	0.012/8	0.012/8	0.010/7	0.008/6
Vehicle test weight wt (lb)	5,018	3,799	3,553	2,449
Vehicle miles per battery (mi)	57,790	50,900	100,000	100,510
Battery capacity (kWh)	44.3	45	53.7	30.6
Design range[a] for 0.25-kW accessory power load (mi)	64.2	80.8	111.1	93.1
Usefulness factor (% of days of year)	91.3	94.6	96.5	96.0
Range penalty for 2.5-kW power load (mi/%)	7.0/14.1	13.7/17.0	22.1/19.9	23.72/25.5
Range increase with regenerative braking (mi/%)				
Recovery factor, RF = 0.5	3.5/5.5	3.9/4.8	5.9/5.3[b]	7.2/7.7
Recovery factor, RF = 0.8	5.8/9.0	6.5/8.0	9.7/8.7	12.1/13.0
Drive motor rated horsepower at 60 mph (HP_R)	25.31	22.55	18.86	14.06
Driving cycle peak motor horsepower (HP_P)	68.76	53.43	48.5	33.57
HP_P/WT	0.0137	0.0141	0.0137	0.0137
Battery-limited HP_p/WT	0.0257	0.0255	0.0148	0.0172
Vehicle length increase over gasoline-powered car[c] (in.)	26.1	18.4	23.1	8.6

[a] SAE J227a Schedule D Driving Cycle.

[b] Not currently feasible without onboard chiller for hydrate ice.

[c] Reference gasoline engine-powered four-passenger car: 1980—1990: (1) HP/WT = 0.015, WT = 2239 lb; (2) HP/WT = 0.030, WT = 2293 lb. 2000: (1) HP/WT = 0.015, WT = 2048 lb; (2) HP/WT = 0.030, WT = 2097 lb. Length = 159.5 in., wheelbase = 97.0 in., height = 53.0 in., width = 60.8 in.

From Donnelly, J. J., Jr., Escher, W. J. D., Greayer, W. C., and Nichols, R. J., *Int. J. Hydrogen Energy*, in press. With permission.

TABLE 11

Total Cost per Mile for Battery-Powered Cars

	Battery			
Characteristics and costs	1985 lead-acid	1985 nickel-zinc	1990 zinc chlorine hydrate	2000 lithium aluminum/ iron sulfide
Capacity (kWh)	44.3	45	53.7	30.6
Design range (mi)	64.2	80.8	111.1	93.1
Efficiencies of battery and charger (%)	70/90	70/90	65/90	90/90
Charge energy at 70% depth of discharge (kWh)	49.2	50.0	64.3	26.4
Charges/100,000 mi[a]	1730	1375	1000	1193
Charge energy cost[a] at 4 ¢/ kWh and 8 ¢/kWh (¢/mi)	3.4/6.8	2.8/5.5	2.6/5.1	1.3/2.5
Total operating cost[b] (¢/mi)	8.9/12.3	8.2/11.0	8.1/10.6	6.8/8.0
Battery cost ($/kWh) (1980 dollars)	40	50—60	50—60	40—60
Vehicle depreciation including salvage (¢/mi)	9.2	9.6—10.6	7.9—8.6	4.9—5.6
Cost of ownership (¢/mi)				
Power cost = 4 ¢/kWh	18.1	17.8—18.8	16.0—16.7	11.6—12.4
Power cost = 8 ¢/kWh	21.5	20.6—21.6	18.5—19.2	12.9—13.6

[a] Assumes vehicle has 100,000-mi lifetime with battery output declining linearly by 20% over the battery cycle life.

[b] Includes costs of charging energy plus 5.5 ¢/mi for repairs, maintenance, tires, accessories, oil, taxes, registration and insurance.[27]

From Donnelly, J. J., Jr., Escher, W. J. D., Greayer, W. C., and Nichols, R. J., *Int. J. Hydrogen Energy,* in press. With permission.

Liquid hydrogen-fueled ICE cars — These vehicles are always the lightest, have the lowest total cost of ownership, and are the most suitable for design ranges above the minimum. They have the lowest source energy consumption for the 1990 model year. Because of their small engine size, they are best suited to higher engine horsepower, if desired. The need for heating of the LH_2 to gasify or heat it provides an opportunity for a simple, lightweight air conditioning system.

Metal alloy hydride-fueled ICE cars — On balance, there is little to choose between the iron-titanium alloy and advanced magnesium alloy hydride systems, except for the better use of coal energy by iron-titanium vehicles. A simple, lightweight air conditioning system is also an inherent capability of the iron-titanium hydride system. Neither hydride system is competitive in the 2000 model year.

Lead-acid battery-powered car — This vehicle has the highest weight, greatest length, highest cost of ownership, lowest usefulness, highest source energy consumption, and is least adaptable to range extension.

Nickel-zinc battery-powered car — This vehicle has high weight, high length, high cost of ownership, and limited range extendability. However, it makes the best use of nuclear energy or electric energy sources in 1990. It accommodates a simple, lightweight air conditioning system option.

Zinc-chloride hydrate battery-powered car — This vehicle is characterized by high minimum range, weight, and length and cost of ownership. It has poor adaptability

TABLE 12

Results of Hydrogen Engine Cycle Analyses

	Naturally aspirated ICE	Direct cylinder injection ICE	
		Compressed hydrogen	Pumped hydrogen
Inlet hydrogen pressure and temperature (psia/°F)	14/80	14/80 14/200	14/$\geq$ −423
Temperature of inlet mix (°F)	108	— —	–
Ideal cylinder charge density (lb/ft³)	0.0482 (mixture)	0.0652 (air) 0.0652 (air)	0.0652 (air)
		0.0671 (corr. for H_2) 0.0671 (corr. for H_2)	0.0671 (corr. for H_2)
H_2 injection pressure and temperature (psia/°F)		1000/697 1000/853	1000/80
Compression work (Btu/lb)			
Air	—	339 339	339
Hydrogen	—	4682 5219	100
Mixture or Total	444	461 476	332
Air pressure and temperature before mixing (psia/°F)	—	534/1288 534/1288	534/1288
Pressure and temperature after compression and mixing (psia/°F)	529/1233	741/1249 766/1306	615/958
Expansion work after constant volume combustion (Btu/lb)	898	904 907	879
Net work (Btu/lb)	454	443 431	547
Thermal efficiency	0.312	0.305 0.280	0.376
Relative thermal efficiency	1.000	0.976 0.897	1.205
Relative displacement/charge mass	1.000	0.718 0.718	0.718
Relative engine mechanical efficiency	1.000	1.062 1.062	1.057
Relative HP/displacement	1.000	1.444 1.327	1.773
Relative total displacement including compressor or pump	1.000	1.001 1.158	0.6204

Note: Compression ratio = 14:1. Inlet air pressure and temperature = 14 psia and 120°F. Fuel/air ratio = 0.029:1; ϕ = 1.0. Compression and expansion efficiencies = 0.75. Two-stage hydrogen compressor with intercooler for 200°F inlet to second stage.

From Donnelly, J. J., Jr., Escher, W. J. D., Greayer, W. C., and Nichols, R. J., *Int. J. Hydrogen Energy,* in press. With permission.

to range extension. However, it makes the best use of nuclear energy or electric energy sources in 1990. It accommodates a simple, lightweight air conditioning system option.

Lithium-aluminum/iron sulfide battery-powered car — This vehicle is lightweight, moderate in length, low in ownership cost, and moderately adaptable to range extension. Its primary advantages are its outstanding values of vehicle energy economy and utilization of all energy resources.

The authors of the study believe that there is a generally uniform degree of optimism in the assumptions underlying the performance of the electric vehicles (EV) considered. However, no substantiation was offered for the LiAl/FeS₂ battery's assumed 20% improvement in all major parameters relative to the 1990 counterpart. Using a more conservative 10% improvement in specific energy and power with all other parameters remaining the same, the LiAl/FeS₂ battery vehicle does not compare favorably with

TABLE 13

Total Cost per Mile for Hydrogen-Powered ICE Cars

Propulsion system	Hydrogen cost ($/lb)	Design range (mi)	Energy cost/mi (¢/mi)	Total operating[a] cost/mi (¢/mi)	Depreciation cost/mi (¢/mi) With hydride salvage	Without hydride salvage	Total cost/mi (¢/mi) With hydride salvage	Without hydride salvage
1985 liquid hydrogen	0.524	85	2.7	8.2	—	4.8	—	13.0
		200	2.7	8.2	—	5.4	—	13.6
1985 FeTi hydride	0.360	85	2.1	7.6	8.1	8.9	15.6	16.4
		200	3.0	8.5	17.7	20.4	26.2	28.9
1990 liquid hydrogen	0.589	85	2.0	7.5	—	4.7	—	12.2
		200	2.1	7.6	—	5.1	—	12.7
1990 FeTi hydride	0.388	85	1.9	7.4	7.3	7.9	14.7	15.3
		200	2.4	7.9	13.3	15.1	21.2	23.0
1990 advanced hydride	0.388	85	2.3	8.3	7.0	7.1	15.3	15.4
		200	2.9	8.9	12.4	12.5	21.3	21.4
2000 liquid hydrogen	0.678	85	2.0	7.5	—	4.2	—	11.7
		200	2.1	7.6	—	4.6	—	12.2
2000 advanced hydride	0.456	85	2.3	8.3	6.1	6.1	14.4	14.4
		200	2.8	8.8	10.5	10.6	19.3	19.4

Note: Assumed salvage values of metal hydrides: FeTi, $2.00/lb: advanced, $0.50/lb.

[a]   Total operating cost per mile = energy cost per mile + taxes, registration, and insurance of 4¢/mi + repairs, maintenance, tires, accessories, and oil cost of 1.5¢/mi for LH_2 and FeTi hydride cars and 2¢/mi for advanced hydride cars.

From Donnelly, J. J., Jr., Escher, W. J. D., Greayer, W. C., and Nichols, R. J., *Int. J. Hydrogen Energy*, in press. With permission.

TABLE 14

1990 Model Year Vehicle Energy Comparisons — Energy Source: Coal

Vehicle type	Design range, R (mi)	Recharge[a] energy	Vehicle energy/mi (kWh/mi)	Source-vehicle input energy efficiency		Source energy per vehicle mile	
				Without storage	With storage[b]	Without storage (kWh/mi)	With storage (kWh/mi)
Hydrogen vehicles[c]							
Liquid hydrogen	85	2.95	0.525				
Pipeline				0.415	0.413;[d] 0.387[e]	1.265	1.271;[d] 1.356[e]
Marine				—	0.387[e]	—	1.356[e]
Rail				—	0.387[e]	—	1.356[e]
Truck				0.394	0.392[e]	1.332	1.339[e]
FeTi hydride (GH$_2$ pipeline)	85	4.14	0.736	0.535	0.529;[f] 0.529[g] 0.532;[h] 0.534[i]	1.376	1.392;[f] 1.392[g] 1.384;[h] 1.379[i]
Advanced hydride (GH$_2$ pipeline)	85	5.13	0.913	0.535	0.529;[f] 0.529[g] 0.532;[h] 0.534[i]	1.706	1.723;[f] 1.725[g] 1.715;[h] 1.709[i]
Battery vehicle							
Zinc chlorine hydrate	111.1	57.8	0.521	0.300	0.291;[j] 0.278[k]	1.735	1.789;[j] 1.872[k]

[a] Recharge energy: hydrogen refill weight in pounds = Cx capacity; C = 0.90 and 0.95 for LH$_2$ in 1985 and later, respectively; C = 0.80 for FeTi hydride in 1985; C = 0.90 for all hydrides in 1990 and 2000. Energy from charger (in kWh) = depth of discharge × capacity/ battery efficiency.

[b] Nonessential energy storage = 10% of directly transmitted energy.

[c] Hydrogen assumed derived from coal gasification because of much higher efficiency than for industrial or residential electrolysis.

[d] LH$_2$ tank.

[e] LH$_2$ tank and truck distribution.

[f] Underground.

[g] Linepack.

[h] Pressure vessel.

[i] Metal hydride.

[j] Batteries.

[k] Electrolytic H$_2$ stored in hydride until used to fuel turbogenerator.

From Donnelly, J. J., Jr., Escher, W. J. D., Greayer, W. C., and Nichols, R. J., *Int. J. Hydrogen Energy*, in press. With permission.

TABLE 15

Comparison of Key Characteristics of Hydrogen- and Battery-Powered Cars

Propulsion system	Design range R. (mi)	Usefulness (% of days)	Curb weight (lb)	Total cost of ownership (¢/mi)	Vehicle input energy (kWh/mi)	Source energy per vehicle mile (kWh/mi) Coal	Nuclear	Wheelbase increase over gasoline ICE car (in.)	Curb weight increase (lb)	Ownership cost increase (¢/mi)	Wheelbase increase (in.)
1985 Model year											
Liquid hydrogen ICE	85	95.1	2220	13.0	0.768	2.059	4.800	3.5	180	0.6	5.0
FeTi hydride ICE	85	95.1	2940	16.4	0.868	1.797	4.173	4.8	2060	12.5	11.2
Lead-acid battery	64.2	91.3	4718	18.1—21.5	0.690	2.300	2.749	26.1	Practically impossible		
Nickel-zinc battery	80.8	94.6	3499	17.8—21.6	0.557	1.856	2.219	18.4	Practically impossible		
1990 model year											
Liquid hydrogen ICE	85	95.1	2100	12.2	0.525	1.265	2.793	2.3	110	0.5	3.3
FeTi hydride ICE	85	95.1	2740	14.7—15.3	0.736	1.376	3.041	3.0	1245	6.4—7.7	6.2
Advanced hydride ICE	85	95.1	2665	15.3—15.4	0.913	1.706	3.773	6.8	1015	6.0	12.5
Zinc chlorine hydrate battery	111.1	96.5	3253	16.0—19.2	0.521	1.735	1.951	23.1	3450		48
2000 model year											
Liquid hydrogen ICE	85	95.1	1895	11.7	0.439	0.994	1.892—1.514	2.0	95	0.5	2.7
Advanced hydride ICE	85	95.1	2340	14.4	0.768	1.372	2.621—2.093	5.5	800	4.9—5.0	10.6
LiAl/FeS$_2$ battery	93.1	96.0	2149	11.6—13.6	0.256	0.852	0.790—0.632	8.6	1570		17

From Donnelly, J. J., Jr., Escher, W. J. D., Greayer, W. C., and Nichols, R. J., *Int. J. Hydrogen Energy,* in press. With permission.

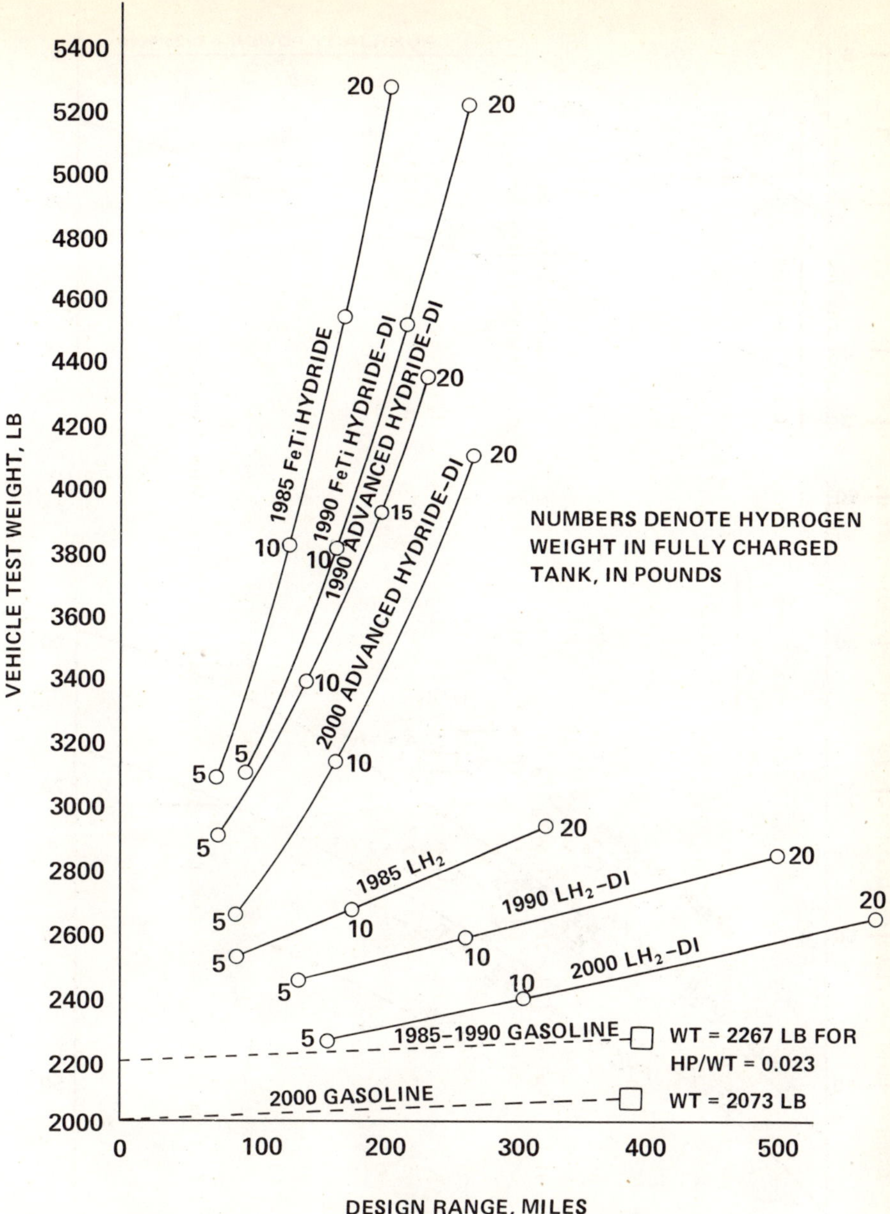

FIGURE 45.   Weight of hydrogen-powered ICE four-passenger cars.[31]

liquid hydrogen or advanced hydride ICE vehicles of the year 2000. The battery system is heavier, longer, more range sensitive, and perhaps more costly to own. However, it retains an advantage in input energy per mile.

Finally, it should be remembered that EVs accommodate regenerative braking, which is a practical advantage for increasing the minimum range by up to 8 to 13%. However, all EVs suffer substantial range loss (14 to 25% at minimum design range) at an accessory power level of 2.5 kW, such as could easily be required when heating is needed.

Another comparative study of hydrogen and battery vehicles is underway under Department of Energy sponsorship. This study, which is even broader and more compre-

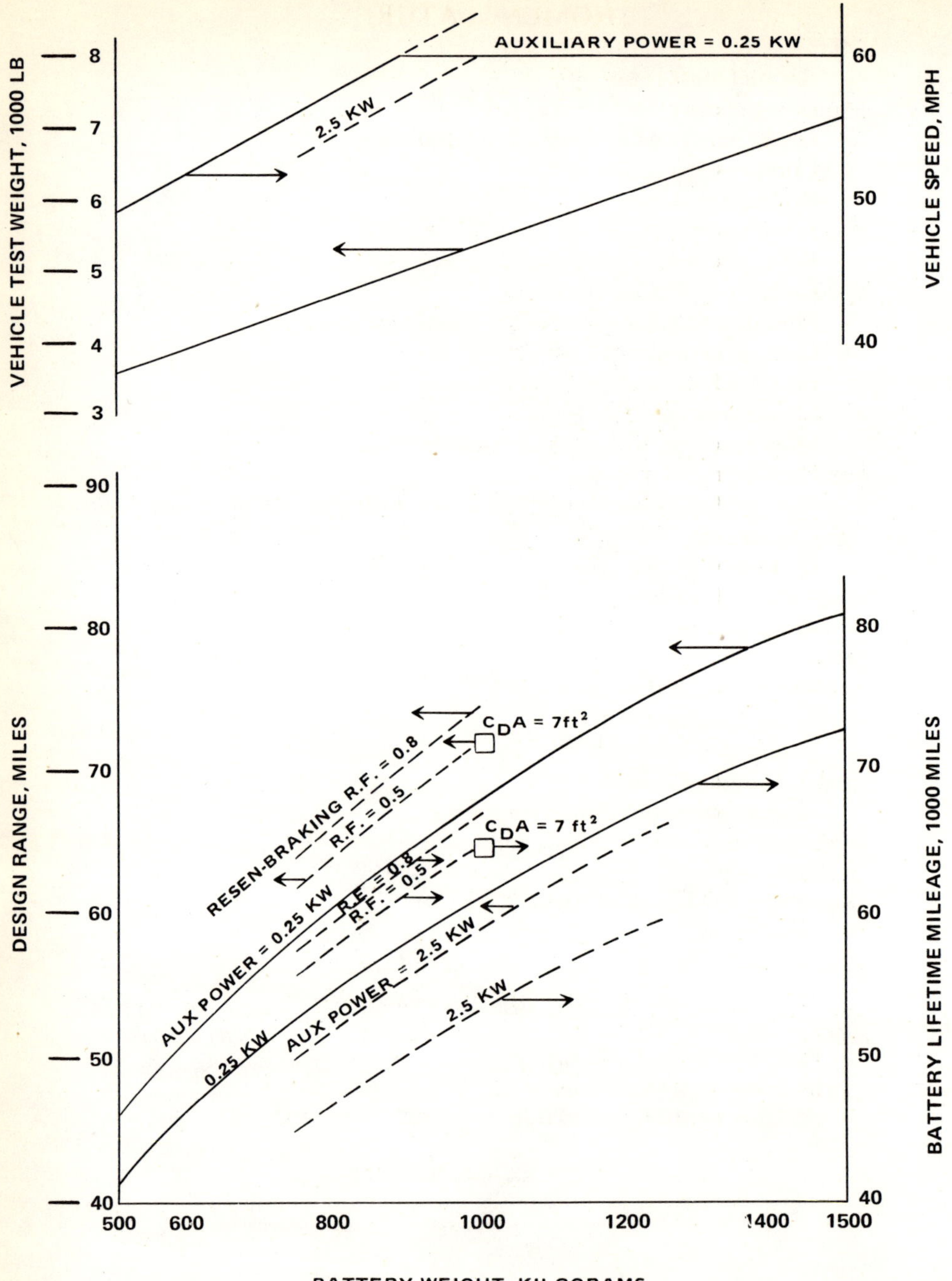

FIGURE 46.   Projected electric vehicle characteristics for 1985 model year using an advanced lead/acid battery.

hensive in scope and manpower resources, is being conducted by the Lawrence Livermore Laboratory[65] in cooperation with the Lawrence Berkley Laboratory, the Brookhaven National Laboratory, Argonne National Laboratory, and the Interplan Corporation. The first-year effort brought out the difficulty in all attempts to project battery and hydride vehicle development. The second-year effort is expected to refine the preliminary results and present a realistic prediction of the prospects for a battery or hydrogen alternative to the gasoline automobile.

NOMENCLATURE

A_f = Air fuel mass ratio
BMEP = Break mean effective pressure
C_dA_f = Drag coefficient and vehicle frontal area
CR = Compression ratio
D_{RR} = Rolling resistance
D_{WR} = Wind resistance
e = Brake thermal efficiency of engine
E_{dt} = Efficiency of drive train
f(V) = Function of velocity in rolling resistance
G_R = Gear ratio (overall)
IMEP = Indicated mean effective pressure
$\dot{M}$ = Mass flow rate
M_f = Mass fraction, hydrogen to hydride mass
M_{MH} = Mass of metal hydride
M_V = Mass of fully loaded vehicle without hydride
N = Engine rpm
P = Brake power of engine
Q = Volume flow rate
R = Range at constant speed
T_d = Tire diameter
v = Volumetric efficiency of engine
V = Road velocity
V_d = Engine displacement
w = Molecular weight
x = Mole fraction or volume fraction
ϕ = Equivalence ratio, stoichiometric air-fuel ratio divided by the actual air-fuel ratio
Δh = Lower heating value of fuel
ϱ = Density
ϱ_o = Air density at reference condition of P_o, T_o.

Subscripts

o = Reference state P_o, T_o for air
gaso = pertains to gasoline engine
H_2 = pertains to hydrogen engine

REFERENCES

1. **Cecil, W.,** On the application of hydrogen gas to produce a moving power in machinery; with a description of an engine which is moved by the pressure of the atmosphere, upon a vacuum caused by explosions of hydrogen gas and atmospheric air, *Trans. Cambridge Philos. Soc.,* 1, 1822.
2. **Erren, R. A. and Hastings-Campbell, W.,** Hydrogen: a commercial fuel for international combustion engines and other purposes, *J. Inst. Fuel,* 6, 277, 1933.
3. **Lichty, L. C.,** *Combustion Engine Processes,* McGraw-Hill, New York, 1967.
4. **Bryant, L.,** personal communication, 1974.
5. **Weil, K. H.,** The hydrogen I.C. engine — its origin and future in the emerging energy-transportation-environment system, No. SAE 729212, in *Proc. 7th Intersociety Energy Conversion Engineering Conf.,* American Chemical Society, Washington, D.C., 1972, 1355.

6. **Ricardo, H. R.,** Further note on fuel research. I., *Proc. Inst. Automob. Eng. London,* 5(18), 327, 1924.

7. **Burstall, A. F.,** Experiments on the behavior of various fuels in a high speed internal combustion engine, *Proc. Inst. Automob. Eng. London,* 22, 358, 1927.

8. **Just, J. S.,** Hydrogen as a substitute fuel, *Gas Oil Power Am. Tech. Rev.,* 326, 1944.

9. **deBoer, P. C. T., McLean, W. J., and Homan, H. S.,** Performance and emissions of hydrogen fueled internal combustion engines, in Proc. Symp. Course Hydrogen Energy Fundamentals , University of Miami, Coral Gables, Fla., 1975.

10. **Oemichen, M.,** Wasserstaff als Motortreib-mittel, in *Verein Deutsche Ingenieur,* Deutsche Kraftfahrtforschung, Verlag GMBH, Berlin, 1942.

11. **King, R. O. and Rand, M.,** The oxidation, decomposition, ignition and detonation of fuel vapors and gases. XXVII. The hydrogen engine, *Can. J. Technol.,* 33, 445, 1955.

12. **King, R. O. Wallace, W. A., and Mahaptra, B.,** The oxidation, decomposition, ignition and detonation of fuel vapors and gases. V. The hydrogen engine and the nuclear theory of ignition, *Can. J. Res. Sect. F.,* 26, 264, 1948.

13. **King, R. O., Hayes, S. V., Allan, A. B., Anderson, R. W. P., and Walker, E. J.,** The Hydrogen engine: combustion knock and related flame velocity, *Trans. E.I.C.,* 2, 143, 1958.

14. **Mulready, R. C.,** Liquid hydrogen engines, in *Technology and Uses of Liquid Hydrogen,* Scott, R. B., Denton, W. H., and Nicholls, C. M., Eds., Pergamon Press, Oxford, 1964.

15. **Crawford, F. H. and Van Vorst, W. D.,** *Engineering Thermodynamics,* Harcourt Brace & World, New York, 1968, chap. 7.

16. **deBoer, P. C. T., McLean, W. J., and Homan, H. S.,** Experimental results with hydrogen fueled internal combustion engines, in Proc. Joint U.S.-Japan Seminar Key Technologies Hydrogen Energy System (NSF/JSPS), Tokyo, 1975.

17. **Messerole, J. S., Jr. and deBoer, P. C. T.,** Reciprocating pump for conversion of liquid hydrogen to high pressure gaseous hydrogen, in Proc. 1st World Hydrogen Energy Conf., Vol. 3, Miami Beach, March 1976.

18. **The Erren combustion cycle,** *Automot. Eng.,* 32(426), August 1942.

19. **Billings, R. E., Baker, N., Lynch, F., and Mackay, D.,** Ignition parameters for hydrogen engines, in Proc. 9th Intersociety Energy Conversion Engineering Conf., San Francisco, 1974.

20. **Van Vorst, W. D. and Finegold, J. G.,** Automotive hydrogen engines and onboard storage methods, in Proc. Hydrogen Energy Fundamentals Symp. Course, University of Miami, Coral Gables, Fla., 1975.

21. **Finegold, J. G., Lynch, F. E., Baker, N., Takahashi, R., and Bush, A. F.,** The UCLA Hydrogen Car: Design, Construction and Performance, Paper SAE 730507, Society of Automotive Engineers, New York, 1973.

22. **Billings, R. E. and Lynch, F. E.,** Performance and Nitric Oxide Control Parameters of the Hydrogen Engine, Publ. No. 73002, Energy Research, Provo, Utah, 1973.

23. **Woolley, R. L.,** Hydrogen engine NO$_x$ control by water induction, paper presented at the NATO/CCMS 4th Int. Symp. Automotive Propulsion Systems, Arlington, Va., April 1977.

24. **Murray, R. G., Schoeppel, R. J., and Gray, C. L.,** The hydrogen engine in perspective, Paper SAE 729216, in *Proc. 7th Intersociety Energy Conversion Engineering Conf.,* American Chemical Society, Washington, D.C., 1972, 1375.

25. **Swain, M. R. and Adt, R. R., Jr.,** The hydrogen air fueled automobile, Paper SAE 729217, in *Proc. 7th Intersociety Energy Conversion Engineering Conf.,* American Chemical Society, Washington, D.C., 1972, 1382.

26. **Adt, R. R., Jr., Hershberger, D. L., Kartage, J., and Swain, M. R.,** The hydrogen-air fueled automobile engine. Part I, in *Proc. 8th Intersociety Energy Conversion Engineering Conference,* 1973, 194.

27. **MacCarley, C. A. and Van Vorst, W. D.,** Electronic fuel injection techniques for hydrogen powered I.C. engines, in Proc. 2nd World Hydrogen Energy Conference (supplement), Zurich, August, 1978.

28. **Homan, H. S., Reynolds, R. K., de Boer, P. C. T., and McLean, W. J.,** Hydrogen-fueled diesel engine without timed injection, *Int. J. Hydrogen Energy,* submitted; **Homan, H. S.,** Ph.D. thesis, Cornell University, Ithaca, N.Y., 1978.

29. **Drexl, K. W., Holzt, H. P., and Gutmann, M.,** Characteristics of a single-cylinder hydrogen-fueled IC engine using various mixture formation methods, paper presented at the NATO/CCMS 4th Int. Symp. Automotive Propulsion Systems, Vol. 2, Session 6, Arlington, Va., April 1977.

30. **Adt, R. R., Swain, M. R., and Pappas, J. M.,** Hydrogen Engine Performance Analysis Project, ERDA Contract E(04-3)-1212, Hawthorne Research and Testing, Inc., First Annual Report, March 1978.

31. **Donnelly, J. J., Jr., Escher, W. J. D., Greayer, W. C., and Nichols, R. J.,** Study of Hydrogen-Powered versus Battery-Powered Automobiles, DOE Report CONS/1101-2, Aerospace Corporation Report No. ATR-77(7385)-1, December 1977; *Int. J. Hydrogen Energy,* submitted.

32. **Burstall, A. F.,** Experiments on the behavior of various fuels in a high speed internal combustion engine, *Proc. IAE,* 22, 358, 1927.

33. **Downs, D., Walsh, A. D., and Wheeler, R. S.,** Knock in the spark-ignition engine, *Proc. R. Soc. London,* 243, 870, 1951.

34. **Woolley, R. L. and Hendrickson, D. L.,** Water induction in hydrogen-powered IC engines, *Int. J. Hydrogen Energy,* 1, 401, 1977.

35. **Anderson, V. R.,** Hydrogen energy in United States Post Office delivery systems, in 2nd World Hydrogen Energy Conf., Zurich, Switzerland, August 1978.

36. **Williams, L. O.,** Hydrogen powered automobiles must use liquid hydrogen, *Cryogenics,* 13, 693, 1973.

37. **Reilly, J. J. and Wiswall, R. H., Jr.,** Iron titanium hydride, *Inorg. Chem.,* 7, 2254, 1968.

38. **Hoffman, K. C., Sinsche, W. E., Wiswall, R. H., Jr., Reilly, J. J., Sheehan, T. V., and Waide, C. H.,** Preprint No. 690232, Society of Automotive Engineers, Warrendale, Pa., 1969.

39. **Woolley, R. L., Beyer, R. B., and Rappley, J.,** Refueling hydrogen transit fleets — data, paper presented at the NATO/CCMS 4th Int. Symp. Automotive Propulsion Systems, Vol. 2, Session 6, Arlington, Va., April 1977.

40. **Strickland, G., Reilly, J. J., and Wiswall, R. H., Jr.,** An engineering scale energy storage reservoir of iron-titanium hydrides, in Proc. Hydrogen Economy Miami Energy (THEME) Conf., University of Miami, Coral Gables, Fla., 1974.

41. **Waide, C. H., Reilly, J. J., and Wiswall, R. H., Jr,** The application of metal hydrides to ground transport, in Proc. Hydrogen Economy Miami Energy (THEME) Conf., University of Miami, Coral Gables, Fla., 1974.

42. **Breshears, R., Cotrill, H., and Rupe, J.,** Partial hydrogen injection into internal combustion engines: effect on emissions and fuel economy, paper presented at the EPA 1st Symp. Low Pollution Power Systems Development, Ann Arbor, Mich., October 1973.

43. **Kester, F. L., Konopka, A. J., and Camara, E.,** Automotive hydrogen storage feasibility: methanol steam reforming, in Proc. 1st World Hydrogen Energy Conf., Miami, Fla., 1976.

44. **Ullman, A. Z. and Van Vorst, W. D.,** Methods of on-board generation of hydrogen for vehicular use, in Proc. 1st World Hydrogen Energy Conf., Miami, Fla., 1976.

45. **Wooley, R. L.,** Design considerations for the Riverside Hydrogen Bus, in 2nd World Hydrogen Energy Conf., Zurich, August 1978.

46. **White, R. A. and Korst, H. H.,** The Determination of Vehicle Drag Contributions from Coast-Down Tests, in SAE Paper No. 720099, Society of Automotive Engineers, Warrendale, Pa., January 1972.

47. **Woolley, R. L.,** Recharging Hydrogen-Fueled Transit Fleets, DOE Report CONS/1293-1, Contract 5]52, Billings Energy Corp., August 1977.

48. **Buchner, H.,** The Hydrogen/Hydride Energy Concept, Daimler Benz, Euratom-Course Hydrogen Energy Concept, Ispra, 1977.

49. **Stewart, W. F., Edeskuty, F. J., Williamson, K. D., Jr., and Lutgen, H. M.,** Operating experiences with a liquid hydrogen fueled vehicle, in *Advances in Cryogenic Engineering,* Vol. 20.

50. **Finegold, J. G. and Van Vorst, W. D.,** Crash test of a liquid hydrogen automobile, in Proc. 1st World Hydrogen Energy Conf., Miami, Fla., 1976.

51. **Peschka, W. and Carpetis, C.,** Cryogenic storage tank for a passenger car, in 1st World Hydrogen Energy Conf., 2B-43, Miami Beach, Fla., 1976.

52. **Crowe, B. J.,** Fuel cells — a survey, NASA SP-5115, NASA, Washington, D.C., 1973.

53. **Marks, C., Rishavy, E. A., and Wyczalek, F. A.,** Electrovan — a fuel cell-powered vehicle, SAE Paper 670176, presented at SAE Automotive Engineer Congr., Detroit, January 1967.

54. **Kordesch, K. V.,** Hydrogen-Air/Lead Battery Hybrid System for Vehicle Propulsion, Abstr. No. 10, Electrochemical Society, 1970.

55. **Kordesch, K. V.,** City Car with H_2-Air Fuel Cell/Lead Battery (One Year Operating Experience), Intersociety Energy Conversion Engineering Conference, Boston, August 1971, 103.

56. **McCormick, B., Depp, S., Srinivasin, S., McBreen, J., Harry, J. L., Voelker, A., Pax, C., Hamilton, D., Kerwin, W., and Baker, B. S.,** Fuel Cell Powered Vehicle Workshop, Division of Transportation and Conservation Research and Technology, Department of Energy, Los Alamos Scientific Laboratory, August 1977.

57. **Billings, R. E.,** A hydrogen-powered mass transit system, 1st World Hydrogen Energy Conf., Miami Beach, Fla., March 1976.

58. **Woolley, R. L.,** Performance of a hydrogen-powered transit vehicle, in Proc. 11th Intersociety Energy Conversion Conf., Lake Tahoe, 1976.

59. **Buchner, H. and Saufferer, H.,** Results of hydride research and the consequences for the development of hydride vehicles, paper presented at the NATO/CCMS 4th Int. Conf. Automotive Propulsion Systems, Vol. 2, Session 6, Arlington, Va., April 1977.

60. **Billings, R. E.,** Hydrogen homestead, in 2nd World Hydrogen Energy Conf., Zurich, August 1978.

61. **Ruckman, J. H., Billings, R. E., Woolley, R. L., Anderson, V. R., Campbell, B. C.,** Progress report on hydrogen production and utilization for community and automotive power, 13th Intersociety Energy Conversion Engineering Conference, August 1978.

62. **Buchner, H. and Saufferer, H.,** Der Wasserstoff/Benzin Mischbetrieb, eine Ubergangaslosung von der Erdol-zur Wasserstoff technologie, in 2nd World Hydrogen Energy Conf., Zurich, August 1978.

63. **Buchner, H. and Saufferer, H.,** Application of low- and high-temperature hydrides: the hydride/hydrogen energy concept, in 2nd World Hydrogen Energy Conf., Zurich, August 1978.

64. **Foster, R. W. and Escher, W. J. D.,** Hydrogen-Fueled Railroad Motive Power Systems—A Feasibility Study, DOE Report CONS/4707-1, U.S. Department of Energy, Washington, D.C., September 1976.

65. **Behrin, E., Bolger, J., Hudson, C. L., O'Connel, L. G., Rubin, B., Schwartz, M. W., Waide, C. H., and Walsh, W. J.,** Energy Storage Systems for Automobile Propulsion, UCLR-52303, Vol. 1 and 2, Lawrence Livermore Laboratory, December 15, 1977.

Chapter 2

Hydrogen-fueled Aircraft

Chapter 2

HYDROGEN-FUELED AIRCRAFT

G. D. Brewer

TABLE OF CONTENTS

2.1. INTRODUCTION

Commercial transport aircraft must be capable of operating anywhere in the world. Obviously, fuel for the aircraft must be available wherever the aircraft are expected to operate. Since petroleum is fast becoming a critically short commodity, there is serious concern that conventional jet fuel may not long be available on an economically acceptable basis.

In the U.S., production of crude oil reached its peak in November 1970. Worldwide, the peak of availability of petroleum is expected to be reached sometime in the period between 1990 and 2000.[1,2] Since commercial aircraft are intended to have an operational lifetime extending 35 to 40 years from start of design, it is timely that serious consideration be given now (1977) to selection of the fuel that can be used in aircraft when Jet A, the petroleum-based fuel currently used in commercial aircraft, is no longer either universally available or competitively priced.

In this chapter, some of the interesting uses which have been found for hydrogen in aeronautics to date are reviewed, and then the potential for its use as a fuel in advanced designs of both supersonic and subsonic commercial transport aircraft is examined. The experience with hydrogen in aeronautics summarized in this brief review provides a basis for confidence that hydrogen can be successfully and safely employed as a fuel in commercial aircraft.

2.2. HYDROGEN IN AERONAUTICS

2.2.1. Airships

The first industrial use of hydrogen was for the inflation of balloons in 1783. The next significant step leading to the use of hydrogen in commercial air transportation occurred in 1900 when the LZ-1, the first rigid airship designed by Count Ferdinand Von Zeppelin, made a successful flight. It used hydrogen gas to provide aerostatic bouyancy. In 1911, the first commercial air operations were started by a German transportation company, Delag (Deutsche Luftschiffahrts-Aktien-Gesellschaft), using five Zeppelin airships. In the next 5 years, before the operations were terminated by World War I, these hydrogen-filled airships made 1600 flights carrying 37,250 passengers without injury.

World War I catalyzed tremendous development activity in airships on both sides. Many hydrogen-filled airships and barrage balloons saw service. The Treaty of Versailles prohibited further development in Germany and stopped manufacture of airships by the Zeppelin company until the U.S. Navy commissioned construction of the airship *Los Angeles,* the LZ-126, as part of reparations. In 1924, the Zeppelin factory at Lake Constance, Germany completed manufacture of the LZ-126, a 2,470,000-ft^3 airship, and delivered it to the U.S. by a transatlantic flight in October 1924. In the service of the U.S. Navy, the *Los Angeles* was inflated using helium. It was flown for a total of 4320 hr and was decommissioned at Lakehurst in 1932. It was dismantled in 1939.

The next airship built by the Zeppelin company, the LZ-127, was the original Graf Zeppelin. It was completed in September 1928 and saw 9 years of continuous, successful service. When decommissioned in 1937, this 3,708,600-ft^3, hydrogen-filled rigid airship had made 590 flights (including 144 ocean crossings, mostly between Germany and South America) and had flown 1,053,391 mi, carrying 13,110 passengers and 235,300 lb of cargo.

The next Zeppelin to be built was the LZ-129, a 7,063,000-ft^3 design christened *Hindenburg.* It was commissioned early in 1936. The airship was 803 ft long and employed the conventional Zeppelin design with 36 longitudinal girders and 15 wire-braced main transverse frames. Hydrogen-filled gas cells were contained within the cloth envelope covering the aluminum framework. Propellers turned by four 110-hp Mercedes-Benz diesel engines provided the *Hindenburg* with a maximum speed of 84 mph. At a cruising speed of 78 mph, the airship had a range of 8750 mi, carrying 50 passengers in spacious accommodations.

In 1936, the first commercial air service across the Atlantic between Germany and the U.S. was initiated with the *Hindenburg* carrying 1002 passengers on ten scheduled round trips. Eastbound crossings averaged 65 hr; westbound, 52 hr. On May 6, 1937, while landing at Lakehurst, the *Hindenburg* burst into flames and was destroyed. Of the 97 persons on board, passengers and crew, 35 lost their lives. In addition, one member of the ground crew was killed when crushed by one of the engines from the airship. These were the first passenger fatalities in the history of commercial airship operation.

The cause of the fire is officially recorded as being due to a discharge of atmospheric electricity in the vicinity of a hydrogen leak. There had been an electrical storm in the vicinity of Lakehurst 2 hr before the arrival of the *Hindenburg,* and the theory is credible. A strong case has also been made for the possibility that the fire was initiated as an act of sabotage by antifascist sympathizers.[3] Elements of a photographic flash bulb were found in the wreckage, and subsequent investigations indicated that a member of the crew, killed in the accident, was involved in suspicious activity both before and during the flight. In any event, the crash of the *Hindenburg* signaled the end of the use of hydrogen as the inflation medium in airships.

The LZ-130, also bearing the name Graf Zeppelin, was a slightly modified sister ship of the *Hindenburg* and was designed to use helium. It was completed and tested (using hydrogen) in September 1938. However, the tense international situation at the time was sufficient grounds for the U.S. to refuse permission to export helium to Germany, so the LZ-130 was inflated with hydrogen for a limited number of exhibition and test flights over Germany. Both the original and the new Graf Zeppelins were dismantled in 1940 to provide aluminum for the Nazi war effort.

2.2.2. NASA-Lewis Flight Research Program

In 1955, a report by Silverstein and Hall of the (then) NASA-Lewis Flight Propulsion Laboratory was published in which the potential of liquid hydrogen as a fuel for use in both subsonic and supersonic aircraft was explored.[4] Among other advantages, the significant improvement in maximum range which theoretically can be realized by using hydrogen was observed.

As a result of this study, an experimental program was initiated to demonstrate the feasibility of burning hydrogen in an aircraft turbojet engine at high altitude. A U.S. Air Force B-57 twin-engine medium bomber was modified as shown in Figure 1 and first flown in 1956. Liquid hydrogen (LH_2) was carried in a tank located under the left wing tip. Gaseous helium was carried in a tank of similar size and shape under the right wing tip for use as a pressurant.

Initially in the program, the hydrogen tank was pressurized to 55 psia with gaseous helium to cause the LH_2 to flow to a heat exchanger where it was vaporized. The heat exchanger was a simple air/hydrogen design where air under flight conditions provided the heat sink to convert the LH_2 to gaseous form.

The regulator which controlled the flow rate of the gaseous hydrogen (GH_2) to the engine was located downstream of the heat exchanger. It was a ratio controller which utilized the metered JP-4 fuel flow from the conventional engine speed control to adjust the flow of hydrogen gas from the heat exchanger. The weight flow rates were adjusted to be inversely proportional to the heats of combustion of the two fuels. When the engine was operating on hydrogen, the flow of JP-4 fuel was recirculated back to its tank.

During the later phases of the program, instead of the pressure feed system, the LH_2 was pumped from the wing tip tank using a specially designed, positive displacement pump located within the tank but shaft driven from outside the tank by a hydraulic motor.[5] The last three flights in the test program were conducted using the pump-fed system.

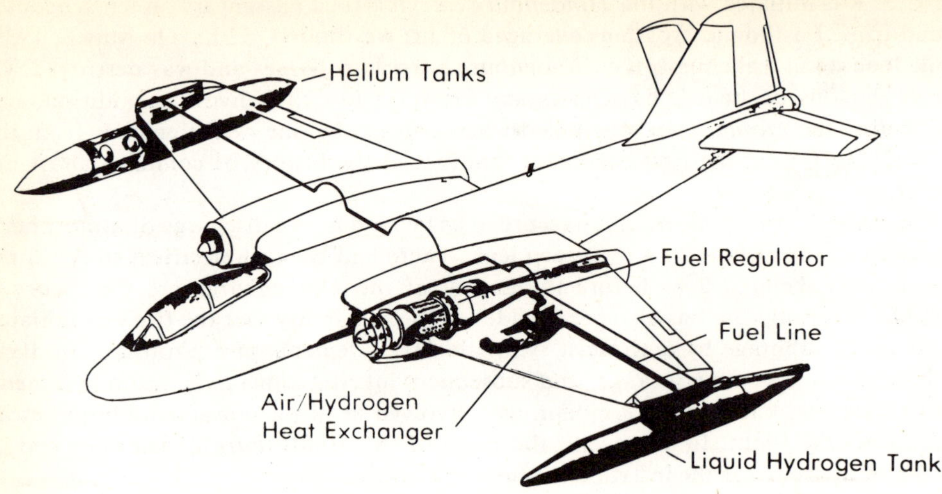

FIGURE 1.   B-57 aircraft used in 1956 NACA test of hydrogen as fuel for aircraft engines.

In the NACA hydrogen flight test program, the converted B-57 aircraft took off and climbed to the altitude and speed specified for the test, typically 50,000 ft and Mach 0.75 over Lake Erie, using conventional JP-4 fuel in both engines. Upon reaching test conditions, the flow of hydrocarbon fuel to the convertible J-65 turbojet engine on the left-hand side was reduced and stopped while the flow of GH_2 was initiated and increased to the required rate. Enough LH_2 was carried in the wing tip tank for about 17 min of operation. When the LH_2 supply was exhausted, JP-4 flow to the engine was restored for the flight back to base. The hydrogen tank and plumbing system were purged with helium before landing. During the NACA hydrogen flight test program, no operational safety problems with the hydrogen fuel system were reported.

2.2.3. The CL-400 Airplane Project

In 1954 to 1955, Lockheed Aircraft Corporation made a series of conceptual design studies of hydrogen fueled aircraft in cooperation with Pratt & Whitney® Aircraft and the Rex Division of AiResearch Corporation. In 1956, the U.S. Air Force awarded a contract to Lockheed's Advanced Development Projects organization, better known as Kelly Johnson's Skunk Works, to build two prototype aircraft which would be capable of cruising at Mach 2.5 at 100,000 ft altitude.

The CL-400 was to carry a two-man crew, a payload of 1500 lb of reconnaissance equipment, and was to be capable of a range of 2200 nmi. A three-view drawing of the airplane configuration is shown in Figure 2. It had a straight wing of low aspect ratio, similar to that of the F-104. A retractable ventral fin was provided to enhance directional stability. The main and nose gears were centerline-mounted, necessitating stabilizing outriggers housed in the nacelles at the wing tips.

The aircraft fuel tanks were contained within the 10-ft diameter fuselage and were segmented into three sections: two main and one sump tank. The forward tank capacity was 17,800 gal, the aft tank 14,240 gal, and the sump tank 4000 gal. The two main tanks were to be maintained at 19 psig, while the sump tank was to be maintained 2 psi lower for fuel transfer. Incorporated in the sump tank was a "well" which housed

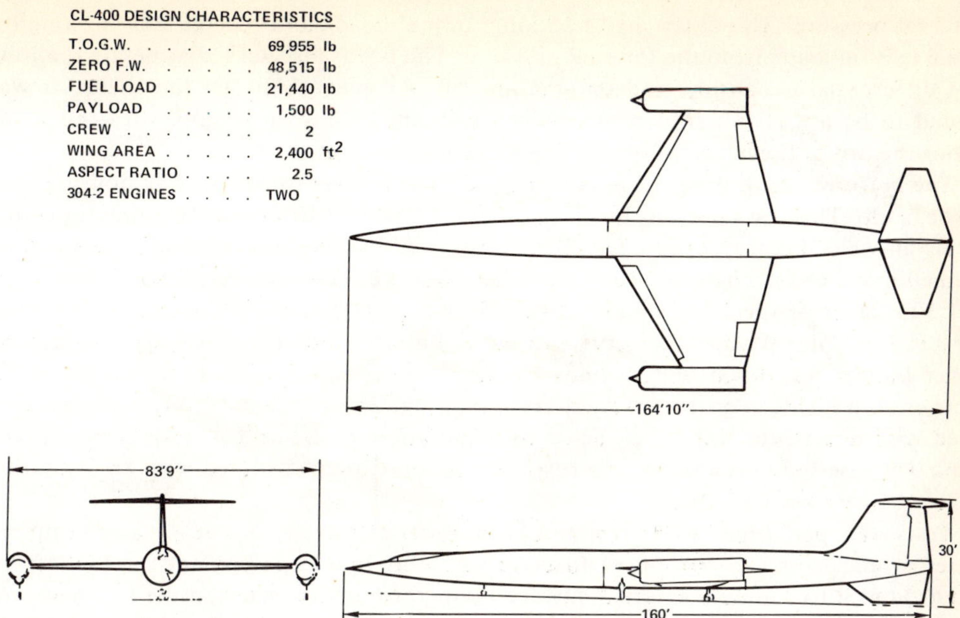

FIGURE 2.   Outline drawing of Lockheed CL-400 hydrogen-fueled reconnaissance aircraft.

the submerged boost pumps. These pumps supplied fuel to the engines at a maximum rate of 385 gal/min each and a pressure of 50 psi. The power plants were located at the wing tips, and it was necessary to transfer the subcooled liquid hydrogen through the wing structure which, in cruise, operated at an average surface temperature of 325°F. Accordingly, the fuel lines were vacuum-jacketed to minimize heat transfer to the LH_2.

The entire propulsion package, including engines, inlets, and nozzles, was to be designed and supplied by Pratt & Whitney® Aircraft. The inlets were a fixed, double-cone, axisymmetric external compression type. The exits were fixed convergent-divergent nozzles. The engine to be used was the Pratt & Whitney® model 304. It will be described subsequently. Each propulsion package weighed 6500 lb and delivered 9500 lb of installed thrust at sea level static with a specific fuel consumption (SFC) of $1.36^{lb/hr}/lb$. At Mach 2.5, 95,000 ft, the thrust was 6100 lb with an SFC of $0.803^{lb/hr}/lb$.

Because of the pioneering nature of the work, Lockheed found it necessary to perform many experiments to verify theory and analysis nearly every step of the way. For example, a small Collins helium cryostat was converted to produce LH_2 in order to gain experience in the production and handling of LH_2. Numerous experiments were conducted to determine the fire and explosion hazards of liquid and gaseous hydrogen. A vacuum-jacketed sight glass was built and installed in a section of a fuel line in a laboratory test set-up to observe the characteristics of flow through a simulated hot wing. This installation was of great value in determining the length of time it took to get stabilized liquid to flow through the system in a representative engine start-up situation.

Many designs of fuel tanks were studied, and several were built and tested. One design which looked promising theoretically but which was found wanting in practice was vacuum insulated with a double aluminum wall separated by ¼ in. of special fiberglass filler. This filler supported the compression load from the outer shell. The inner shell had circumferential beads and reacted only to hoop tension loads from the internal pressure. The outer shell had longitudinal beads and reacted only to longitudinal tension loads from the internal pressure. The beads also served as bellows, allowing differential expansion and contraction to take place between the shells. It was found to be a difficult job to maintain a vacuum in this large tank with the welds along the tips of the beads where flexing was concentrated.

The preferred tank design was constructed in half-scale to permit fuel boil-off tests to be made. This test tank was fabricated from 0.050-in. thick 6061-T6 aluminum alloy using all-welded construction. The diameter was 50 in., the length 85 in., and the tank had elliptical ends. The insulation consisted of a 1-in. layer of styrofoam followed by a 1.5-in. layer of special fiberglass. Each layer of insulation was covered with a wrap of 0.005-in. thick Mylar® to serve as a vapor barrier. The styrofoam layer within the inner barrier was filled with helium gas and the special fiberglass within the outer barrier with nitrogen gas, to prevent cryopumping. The entire assembly was then covered with aluminum foil which acted as a radiation shield and a static ground. The tank was inserted into a half-scale fuselage section which was lined with an insulation layer of fiberglass 1-in. thick.

Tests were performed to determine LH_2 boil-off characteristics at elevated temperatures by enclosing the entire tank-fuselage assembly in a large box and circulating hot air around it to simulate external surface temperature. Corrected to the full-scale aircraft in performance of its design mission, the boil-off was found to be about 3% of the full fuel load. Since about two thirds of this was required to maintain tank pressure, the boil-off loss was only about 1%.

A full-scale, 4000-gal sump tank of this basic design was also built so that a reasonably complete aircraft fuel system could be ground-tested with the Pratt & Whitney® engine.

In spite of the success in developing practical solutions to the problems encountered with handling the cryogenic liquid, the CL-400 aircraft was never built. In 1957, the program was terminated by mutual agreement between the Air Force, Lockheed, and Pratt & Whitney®. There were two basic reasons for this action. First, the airplane was marginal in range capability and did not show potential for significant improvement. This was due partly to the low supersonic lift to drag ratio which could be achieved resulting from the combination of a relatively small wing and the large fuselage needed to contain the low-density LH_2. It was also partly due to the performance which could be obtained with the Model 304 engine. Although at the time the 304 engine represented a significant technical achievement, compared with performance which was to be demonstrated later it was both heavy and relatively inefficient.

A second reason for termination of the CL-400 program was the logistics problem posed by the fuel. The aircraft was intended for reconnaissance purposes which would require flights to many different locations around the world. LH_2 would therefore have to be available at these locations in order for the aircraft to fly the return trip. Since concurrent studies at Lockheed had shown that the desired mission could be performed using a more conventional hydrocarbon fuel, although with a small compromise in cruise altitude, there was little incentive to continue development of the CL-400. The alternate design which was subsequently pursued led to the Skunk Works Blackbird program, the Lockheed SR-71/YF-12 aircraft which used the Pratt & Whitney® J-58 engine.

The CL-400 design and development program showed that it was entirely feasible to build a hydrogen fueled airplane. LH_2 could be handled, with proper care and procedures, as easily and safely as hydrocarbon fuel. A great deal of technology was developed which later proved useful in the U.S. space program. Development data on handling of LH_2, tank construction, and materials were turned over to the Convair

Division of General Dynamics Corporation for their use in the Centaur program, the highly efficient upper stage of a workhorse launch vehicle which is still in service.

2.2.4. Early Turbojet Engine Development

Early in 1956, at the same time the U.S. Air Force contracted with Lockheed for development of the CL-400 airplane, the Pratt & Whitney® Aircraft Division of United Aircraft Corporation (now United Technologies, Inc.) was awarded a contract to investigate the feasibility of using LH_2 as a fuel in aircraft engines. The work at Pratt & Whitney® covered a broad spectrum from applied research efforts, such as heat transfer measurement and materials investigation, to development testing of a J57 engine modified to operate on LH_2 and including the design, construction, and test of a new design of engine, the model 304.

Conversion of the J57 to operate on LH_2 was accomplished in 5 months, and first tests were performed in the fall of 1956. LH_2 was supplied to a single-stage, engine-driven centrifugal fuel pump with a slight net positive suction head to prevent cavitation. To insure that the hydrogen was injected into the combustion chamber as a gas, it was passed through a counterflow heat exchanger where it was warmed by air bled from the compressor discharge. An axial, tube-type injection system was used after early tests showed that it provided acceptable burner-can discharge temperature profiles and that good combustion efficiencies could be obtained with combustor cans as short as about one fourth the length of those required for hydrocarbon fuels.[6]

The work with the J57 showed that conventional jet engines could be readily adapted to use LH_2 fuel. It was desired, however, to provide better performance than that which could be obtained from modifying an existing engine. After examining many possible cycles, the Hydrogen Expander cycle, shown schematically in Figure 3, was selected for experimental evaluation.[6] It was a unique cycle developed specifically to take advantage of the properties of hydrogen and to meet the performance requirements of the CL-400 airplane. In this cycle, the hydrogen fuel was pumped to 600 psi by an engine-driven centrifugal pump, from which it passed through a heat exchanger where it was heated to a temperature of about 1800°R. The hot hydrogen gas was then expanded through 18 turbine stages and divided into two sections of 6 and 12 stages each. At a hydrogen flow of about 4 lb/sec, the turbine produced 12,000 hp at 25,000 rpm. The shaft power passed through a 4:1 reduction gear to drive a five-stage fan to provide a maximum corrected airflow of 542 lb/sec.

To provide the best thrust to weight ratio possible at cruise conditions, the engine was derated at takeoff, with airflow being limited by control of turbine power output. Above the altitude at which maximum corrected airflow could be delivered with full design horsepower, turbine rpm was controlled.

Hydrogen gas from the turbine exhaust passed through injectors into an annular

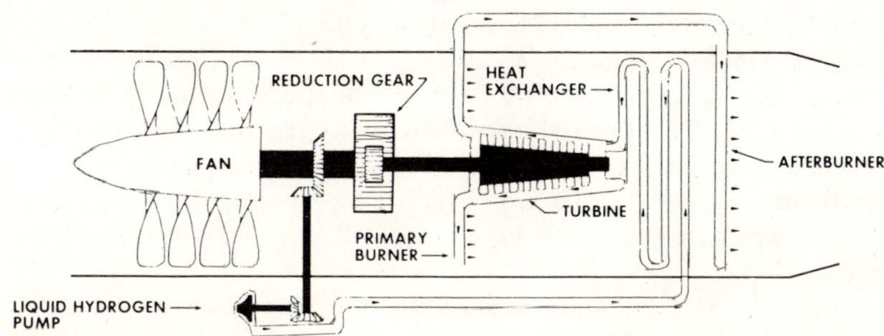

FIGURE 3.   Schematic cycle diagram of the Pratt & Whitney® model 304 engine.

combustion chamber where it mixed with air discharged from the fan and burned at about 2000°R. The combustion gases then flowed through the heat exchanger to provide the energy required by the hydrogen to drive the turbine. The quantity of hydrogen burned in the combustor was controlled to limit the temperature rise of the hydrogen gas as it circulated through the heat exchanger on its way to the turbine. Excess hydrogen, not burned upstream of the heat exchanger, was bypassed to an afterburner just downstream of the heat exchanger. A convergent-divergent nozzle completed the model 304 engine cycle.

The first demonstration test of a complete 304 engine was accomplished in September 1957. A total of 25 hr of sea level engine testing was performed, and all performance predictions were confirmed. Termination of the CL-400 airplane program resulted in cancellation of the Model 304 engine development. Efforts at Pratt & Whitney® relative to hydrogen were redirected to development of the RL10 hydrogen-oxygen liquid rocket engine for use in the U.S. space program.

2.2.5. The U.S. Space Program

The first successful launch of a space vehicle propelled by a liquid hydrogen/liquid oxygen rocket engine took place at Cape Kennedy on November 27, 1963. The launch vehicle was an Atlas-Centaur built by the General Dynamics Corporation Convair Division. The kerosene (RP-1)/liquid oxygen-fueled Atlas stage boosted the Centaur upper stage which employed two of the RL10 rocket engines developed by Pratt & Whitney® Aircraft.

The RL10 used the same hydrogen expander cycle to drive its turbopumps as was successfully demonstrated with the model 304 turbojet engine.[6] In the case of the rocket engine, LH_2, circulated through the cooling jacket of the combustion chamber and nozzle assembly, picked up enough heat so that, when expanded through the turbine, it provided the power to pump both propellants to the pressure required to force them through the injector system into the combustion chamber. The high specific heat and the low critical temperature and pressure of hydrogen made it possible to obtain 800 hp with a two-stage aluminum turbine having a turbine inlet temperature of 135°F below 0 and a flow rate of only 4 lb/sec. In addition, it was found unnecessary to use any conventional lubricants in the turbopump assembly. A small jet of hydrogen gas was used to remove the heat of friction of bearings and gears.

Several other rocket engine manufacturers in the U.S. were concurrently involved in development of designs using LH_2. The General Electric Company, Rocketdyne Division of North American Aviation (now Rockwell International, Inc.) and the Aerojet General Corporation were among the leaders. Of the designs developed by these companies, the Rocketdyne J2 engine, shown in Figure 4, is an example which has been eminently successful. It was used in both the second and third stages of the Saturn V launch vehicle for the Apollo program. There are five J2s clustered in the Saturn V second stage (called the S-II stage)and one J2 in the third or upper stage (called the S-IVB). The S-II stage has a capacity of about 160,000 lb of LH_2, and the S-IVB stage holds approximately 43,000 lb. In all of the launches of the Apollo program there has never been a failure of one of the hydrogen-fueled rocket engines. This remarkable record is at least partly due to the fact that hydrogen is a tractable fuel and, as stated by Mulready,[6] "The use of hydrogen as the fuel for these various types of engines has in each case proved to be a major ally in achieving high performance and long life."

A tremendous body of experience in handling LH_2 has been gained as a result of the development and use of liquid rocket engines in the space program. The experience includes manufacture and liquefaction of the hydrogen, as well as transportation, stor-

FIGURE 4.   Rocketdyne J2 rocket engine.

age, pipelining, instrumentation, design practices, operational use, and safety procedures.

At the same time that chemical rocket systems were being developed for the earth orbital and cislunar projects, a parallel effort was underway to develop nuclear rocket stages for longer range space missions. Because of the importance of minimizing molecular weight of the exhaust products in a rocket-type reaction to achieve high performance, hydrogen was used as the working fluid in the nuclear program. In addition to its molecular weight of two being close to ideal for the exhaust gas, the very high specific heat of hydrogen was important in enabling it to be used effectively as a coolant for the structure of the reactor and nozzle.

The U.S. nuclear rocket program was divided into three phases: ROVER, NERVA, and RIFT. The Los Alamos Scientific Laboratory had overall responsibility for design of the nuclear reactor and for feasibility testing. Under ROVER, the initial testing of nuclear reactors as an energy source for rocket propulsion was performed with reactors named KIWI. These were reactors designed specifically for development testing and were not intended for flight (hence, use of the name of the flightless New Zealand bird). The first reactor test was carried out at the Nevada Test Site of the U.S. Atomic Energy Commission in July 1959. A total of 12 reactors were built and tested as part of the ROVER phase of the program.

Development of a nuclear rocket engine in a flyable configuration was undertaken under the NERVA (Nuclear Engine Rocket Vehicle Application) phase. Aerojet General Corporation was the prime contractor for the complete engine, with Westinghouse Electric Corporation having the contract for design of the reactor itself. The third phase was development of a flight vehicle in which the NERVA engine could be tested under operational conditions. This was RIFT (Reactor In Flight Test). Lockheed Missiles and Space Company was prime contractor.

A comprehensive facility for handling liquid hydrogen and for testing nuclear rocket engines was developed at Jackass Flats, the Nevada Test Site of the U.S. Atomic Energy Commission.[7] In addition, impressive hydrogen facilities were built at the plants of the contractors involved in the program.

The nuclear rocket development program was gradually scaled down and eventually terminated in 1972, primarily for economic reasons. In a period when federal budget cutbacks were necessary, it was felt that the nuclear rocket program, being developed for long-range space projects, was one of the items which could deleted in favor of more immediate needs. The direct results of the program were the successful tests of the KIWI, Phoebus, and Pee Wee reactors demonstrating the feasibility of the concept; significant progress in development of a NERVA engine; and, perhaps most important, development of technology and experience in handling hydrogen which provided a basis for programs of the future. Over 37 million gal of LH_2 were handled at the Nevada test site alone, with no serious injuries due to hydrogen incidents.

2.2.6. Hypersonic Aircraft Studies

Another category of use for hydrogen which has been the subject of serious study and analysis is hypersonic aircraft. Starting in 1965, when investigations led to realization that the lifting body shape might offer significant advantages for hydrogen fueled, hypersonic aircraft, studies have continued to explore their potential. Use of hydrogen as fuel in hypersonic aircraft is a virtual necessity because of the need for cooling aircraft and engine structures. Hydrocarbon fuels do not offer sufficient heat sink capacity.

Aircraft capable of speeds of Mach 5 and above are of interest for several reasons. For passengers, the prospect of flying long distances in very short times is attractive, particularly so if the quicker trip can be provided without significant cost increase. This possibility has been indicated in numerous studies which have been made, (e.g., by Peterson and Waters[9] and Becker and Kirkham[10]).

Another potential advantage of hypersonic aircraft is the possibility that they can reduce the sonic boom overpressure to the point where overland flight can be permitted. This could be possible primarily as a result of the high cruise altitude for such vehicles and also because of the more advantageous shape of the higher speed aircraft.

The possibility also exists that hypersonic aircraft can be used as air breathing, reusable launch vehicles for space missions. They could carry vehicles to be launched into space in piggyback fashion and thus serve the purpose of reusable launch platforms to reduce the cost of space missions.[11]

TABLE 1

Effect of Hydrogen Properties on Airplane Design

Property	Effect (relative to Jet A fuel)
High heating value	Fuel weight $\div$ 2.8
High specific heat	Fuel cool engine hot parts
	High TIT or OPR
	Reduced SFC
	Further weight saving
Low density	Lesser weight of fuel requires about 3.78 times more volume; this leads to
	Lower L/D
	Low wing loading
	Higher cruise altitude
	High power loading
Cryogenic	Requires
	Airtight insulation system
	Heavy tank and fuel system
	Special tank filling and venting procedures
	Constant absolute tank pressure to minimize boil-off

The characteristics of hydrogen which make these hypersonic vehicle applications attractive are the same as previously discussed: high heat of combustion, high specific heat, low molecular weight, and clean burning to minimize pollution of the atmosphere.

2.3. THE POTENTIAL OF HYDROGEN AS A FUEL FOR TRANSPORT AIRCRAFT

In this section the desirability of using LH_2 as fuel for future designs of supersonic and subsonic commercial transport aircraft is examined on a qualitative basis. To make a quantitative assessment of the net advantage/disadvantage which would accrue from using LH_2 in transport aircraft, it is necessary to perform a detailed design study, comparing characteristics of vehicles designed to use LH_2 fuel with comparable aircraft designed to use Jet A fuel. This kind of comparison is made in the following section. First, the principal characteristics of hydrogen are examined to determine the gross trends which might be expected from using this interesting fuel.

Table 1 lists four properties of hydrogen in the left column and, in the column on the right, an indication of the effect these properties would be expected to have on aircraft design, relative to conventional jet fuel. For convenient reference, properties of Jet A and liquid hydrogen which are significant in aircraft application are listed in Table 2.

The ratio of the heats of combustion shown in Table 2 for Jet A fuel and hydrogen favors LH_2 by a factor of 2.8. Accordingly, as indicated in Table 1, the specific fuel consumption (SFC) of a hydrogen-fueled engine would therefore be less by a factor of 2.8, all other considerations remaining equal. In other words, it will take less fuel, by a factor of 2.8, with LH_2 to produce a given level of thrust, compared to that required with Jet A.

The high specific heat of LH_2 permits the fuel to be used as a heat sink so that engine hot parts can be cooled efficiently with less bleed air from the engine cycle, as compared to that required with conventionally fueled aircraft. There are several ways to take advantage of this capability. One which has been proposed is to employ a

TABLE 2

Comparison of Properties of Jet A and Liquid Hydrogen

	Jet A	Hydrogen
Nominal composition	$CH_{1.93}$	H_2
Molecular weight	≈ 170	2.016
Heat of combustion (Btu/lb)	18,400	51,590
Liquid density (lb/ft³) at 50°F	51.6	4.43[a]
Boiling point (°F) at 1 atm	339—513	−423
Freezing point (°F)	−58	−434
Specific heat (Btu/lb °F)	0.47	2.29
Heat of vaporization (Btu/lb)	127	193

[a] At boiling point.

secondary fluid in a closed loop with a heat exchanger. The coolant would be circulated through the high-pressure turbine section to affect the cooling required, then pass through the heat exchanger to transfer its excess heat to the hydrogen before the fuel is injected into the combustor. In this way, the heat is regeneratively used in the engine, higher turbine inlet temperatures (TIT) are possible because of the cooling, and higher overall pressure ratios (OPR) are permitted. The result would be a smaller, lighter, more efficient engine to provide a given thrust output at an even lower SFC than was calculated by ratioing the heats of combustion. Estimates have been made that an additional 5 to 10% improvement in SFC might be realized as a result of (a) not having to bleed air from the engine cycle to do the cooling, (b) increasing the efficiency of the turbine by eliminating the discharge of cooling air from the vanes and rotor blades, (c) returning the heat to the combustion process, and (d) designing to a higher TIT and OPR.

Additional propulsion benefits resulting from use of hydrogen in aircraft engines stem from characteristics of the fuel not listed in the table. On the basis of operational experience with gas turbine engines fueled with natural gas at gas pipeline booster stations, it has been found that engine life is increased about 25% and that engine maintenance requirements are reduced by a similar percentage, compared with identical engines fueled with Jet A. Hydrogen offers even more improvement in these areas than does natural gas (methane). These desirable attributes result from the fact the fuel is clean and free of impurities to either erode or corrode and that it is injected into the combustor in gaseous form so that very rapid diffusion and mixing occur. These latter characteristics promote complete and rapid combustion and, therefore, provide a uniform temperature profile which minimizes thermal stresses in the combustor. The low emissivity of the hydrogen/air flame further reduces metal temperatures for a given average temperature of combustion products so that combustor liners and turbine vanes and blades are subjected to less rigorous operating conditions for a comparable level of performance.

The last two properties of LH_2 listed in Table 1 represent adverse effects on airplane design. The low density of LH_2, compared to Jet A fuel, means that even with the potential saving in fuel weight represented by the reduction in SFC, up to 3.78 times more volume must be provided to contain it. The result is that hydrogen-fueled aircraft must have a different tankage arrangement than conventionally fueled designs, and the result will be higher drag for the aircraft.

LH_2-fueled aircraft can be expected to have low wing loading at takeoff. This results from the fact that there is less fuel weight to burn during the flight. Since the aircraft will be constrained to land in the same distance, wing loading at landing must be equivalent to existing practice.

As a result of the lower average wing loading during cruise, LH_2-fueled aircraft will generally want to cruise at a higher altitude to achieve a maximum lift to drag ratio (L/D). All other things being equal, it would be expected that they would require relatively higher thrust to weight ratio, consistent with the higher cruise capability.

Applying this qualitative analysis to supersonic transport aircraft, it appears that the same trends should hold. Moreover, it follows that sonic boom intensity should be reduced as a result of the higher cruise altitude and hydrogen-fueled aircraft will be lighter, longer, and have smaller wings. The combination of these effects can be expected to make a significant different in sonic boom levels between Jet A and LH_2 fueled supersonic transport aircraft.

A synergistic outfall of the combination of low wing loading at takeoff and high power loading is that LH_2-fueled aircraft may be capable of takeoff with engines throttled, thus reducing noise. Hydrogen-fueled aircraft should permit new levels of quiet to be established around airports.

Continuing with the qualitative analysis of the items listed in Table 1, the fact that LH_2 is cryogenic will serve to complicate both the airplane design and the refueling procedures. The $-423°F$ temperature of LH_2 requires that there be an insulation system applied to the tanks and fuel lines which minimizes heat leak into the fuel, thereby limiting fuel boil-off, and which also prevents contact of air with any supercooled surfaces that would result either in liquefaction or freezing of the air. Design of the fuel tank supporting or attaching structure is complicated by the same necessity to minimize heat leaks to the fuel.

The requirement for a cryogenically insulated system does not, in itself, pose an insurmountable problem; neither does the fact it must be flight-weight. There are numerous examples of flight-weight, insulated structural systems to contain liquid hydrogen in the U.S. space program. These, however, are all single-use, short-duration flights. The aircraft problem lies in designing such a system so it is capable of withstanding the shocks and stresses of thousands of landings and associated flight loads and so it will retain its thermodynamic effectiveness through the 15- or 20-year useful life of the airplane. During that useful life, it must be accessible for routine inspection, maintenance, and/or replacement.

Special procedures must be followed in filling a tank with LH_2. All air must be removed before LH_2 is introduced; otherwise, it will solidify and plug a line or freeze a valve. GH_2 or helium must be used in the final purging operation because these are the only gases which will not freeze on contact with LH_2. A vent system must be provided in the aircraft to permit safe release of GH_2 during the filling operation (and in the event of an extended ground-hold operation when the aircraft tanks are filled but takeoff is delayed) and during flight.

Finally, a special problem posed by use of cryogenic LH_2 for aircraft fuel involves providing for a different means of tank pressurization. With conventional fuel, air is allowed to leave or enter the tanks as the airplane climbs or descends along its flight path, in order to equalize internal and external pressures, thereby minimizing structural weight of the tanks. This, of course, cannot be permitted with LH_2 fuel for two fundamental reasons. First, any air introduced into the LH_2 tank would immediately freeze solid; second, pressure in the hydrogen tank cannot be allowed to fluctuate greatly. This would lead to large losses of LH_2 through boil-off.

For example, assume the LH_2 is tanked as a saturated liquid. As the aircraft climbs to cruise altitude, if tank pressure is allowed to decrease in order to maintain a constant differential with external conditions, the LH_2 would boil until the temperature of the fuel reduced to the point where its temperature and pressure would once again be in equilibrium, i.e., it would again be a saturated liquid (refer to a hydrogen temperature-entropy diagram for clarification, see Volume III of this series). Obviously, much of the GH_2 boiled off in the process of seeking equilibrium at the lower pressure would be lost through venting.

On the other hand, as the aircraft descends, if tank pressure is increased to maintain constant gauge pressure, the liquid would become supercooled.It would stop boiling until heat leak into the tank raised the temperature of the liquid to the point where equilibrium would once more be achieved at the new pressure level. Special means would have to be provided to keep the tank pressurized during descent. As just noted, only GH_2 or helium could be used.

For these reasons, the concept of designing aircraft LH_2 tanks for constant gauge pressure seems impractical. A better approach is to design for constant absolute pressure in the tanks, thereby minimizing LH_2 temperature fluctuations and boil-off losses.

From this qualitative examination it is clear that, while use of LH_2 as fuel in aircraft can provide many advantages, there are also some disadvantages which will be encountered. Most of the disadvantages are design type problems which can be solved by exercising engineering ingenuity and through development effort. There are also problems of establishing procedures which need to be followed for safe and reliable operation. None are considered insurmountable nor of the type that require invention or significant state-of-the-art advances to resolve. On balance, the potential advantages would seem to far outweigh the recognized disadvantages.

2.4. TRANSPORT AIRCRAFT DESIGNS

Based on the favorable prospects for hydrogen resulting from the foregoing qualitative analysis, early in 1972 some quantitative design studies were undertaken at Lockheed-California Company. In 1973 a program was initiated at the National Aeronautics and Space Administration to explore the subject more carefully. As a result of this work, the potential for hydrogen fuel has now been examined in advanced design versions of both supersonic and subsonic aircraft.

Essentially, the objectives of the studies in both speed regimes were identical. They were to assess the feasibility of using hydrogen fuel in commercial transport aircraft, to determine its advantages and/or disadvantages relative to conventional Jet A fuel, to identify problems and technology requirements associated with use of LH_2 and to outline a plan for development of the required technology. The basic guidelines which were used in the studies are listed in Table 3. A brief examination of these guidelines will enhance understanding and appreciation of the results which were obtained.

In order to permit focusing the effort on work related to the primary objectives, it was mandated that LH_2 was to be considered available in storage at the airport. NASA funded other studies to investigate processes, energy requirements, and costs involved in manufacture, transmission, and liquefaction of hydrogen. This work was done by the Institute of Gas Technology, Chicago, and the Linde Division of Union Carbide Corporation, Tonawanda, New York.

For purposes of calculating direct operating costs (DOC) of the aircraft, baseline prices were specified for both LH_2 and Jet A fuel. Variations from these baseline prices were then considered to determine the sensitivity of DOC to fuel cost and to permit comparison of relative DOCs for the aircraft for actual fuel prices which might be established in the future.

TABLE 3

Basic Guidelines for LH₂ Aircraft Studies

Liquid hydrogen assumed to be available at airports
Initial operation capability, 1990—1995
Advanced technologies
    Control configured aircraft
    Composite materials
    Configuration:
        Subsonic — supercritical aerodynamics
        Supersonic — NASA arrow-wing
    1985 State-of-the-art propulsion technology
Direct operating costs
    1967 ATA equations
Design LH₂ and Jet A fueled aircraft to perform same mission, based on same operating requirements:
    Runway length
    Noise limitation
    Fuel reserves
    Certification standards

Initial operational capability (IOC) in the early 1990s was stated as a goal. The feasibility of achieving that goal was to be determined in each study for the two classes of transport aircraft, supersonic and subsonic.

As a result of the IOC being specified 15 years or more into the future, it was further mandated that the aircraft designs should incorporate advanced technologies which are currently under development and which show promise of improving the performance of transport aircraft of the future, regardless of the fuel used. The items listed in Table 3 were included. A control configured aircraft uses an active control system and so is not constrained to meet conventional requirements of static stability, thus serving to minimize size of the control surfaces. Other potential benefits generally associated with use of an active control system on aircraft of the types studied are flutter suppression and load control to minimize structural design loads and dynamic loads experienced by the passengers. Although these latter benefits are real, it was felt that accurate quantitative definition of their implications in the subject aircraft designs was beyond the scope of the work and not fundamental to the primary objective of comparing hydrogen-fueled aircraft designs with equivalent, conventionally fueled versions. Accordingly, evaluation of benefits from these latter effects was not included.

Use of composite materials in aircraft structure has been the subject of serious study for several years. The primary objectives are to reduce weight, improve stiffness, and provide smooth aerodynamic surfaces, while in some instances also providing reduced maintenance and lower cost. For the subject work, an assessment was made of the state-of-the art forecast for the mid-1980s, the period in which design freeze would have to occur in order for aircraft to become operational in 1990 to 1995. The resulting level of materials technology was incorporated into the structural designs and reflected in weight and cost estimates of the subject aircraft.

In similar fashion, concepts of aircraft configurations for both speed regimes were employed which reflected the results of significant analytical effort and wind tunnel testing. The subsonic designs employed the supercritical aerodynamic concepts developed by NASA in recent years and which are now being incorporated in advanced design aircraft. The benefits to be realized through the use of the specially shaped airfoil sections are that they permit use of thicker wings for a given cruise speed, thereby deriving structural advantages or which, alternatively, can be used to minimize drag at a given wing thickness by delaying onset of the drag rise associated with high subsonic speeds.

Use of the NASA arrow-wing planform was specified for the supersonic transport design study because of attractive aerodynamic potential at both low speeds and also at the cruise condition. These benefits were determined analytically and verified by wind tunnel testing at Langley Research Center.

Turbofan engines based on advanced component technology were synthesized parametrically for both fuels, LH_2 and Jet A. Component efficiencies and capabilities which formed the basis for the generation of the engine cycle characteristics and performance data were based on extrapolation of observed development trends, modified by judgment involving awareness of trade-offs to afford maximum benefit and knowledge of limits of practicality and tempered by published predictions of the major engine companies.

Parenthetically, it should be noted that the engine characteristics developed for these studies are not unique to the results which were obtained. It is felt that the engine characteristics developed in these studies for both fuels adequately represent advanced capability propulsion units so that valid comparisons can be drawn between performance of the competing aircraft. Engine designs developed specifically for the applications will undoubtedly differ in many respects; however, the relative performance of engines designed for the two fuels should reasonably approximate the differences reflected herein.

Direct operating cost (DOC) was used as a principal selection criterion in picking preferred aircraft designs during the parametric evaluations. As noted in Table 3, DOC was calculated based on the 1967 Air Transport Association (ATA) formula and specifications.[12] It is recognized that the ATA equations and the related bases for deriving input values are not necessarily equitable with actual airline experience in evaluating DOC. The fact is that no two airlines seem to agree exactly on definition of an equitable method of calculating DOC. The 1967 ATA equations were used to calculate values that are considered representative of the DOC parameter for purposes of comparing the capability of one airplane design with another. The absolute values of costs so calculated may not necessarily have major significance.

The final basic guideline listed in Table 3 was one of the most significant. This was that comparisons of LH_2- and Jet A-fueled aircraft would be made only if they were designed to identical standards. Both aircraft must be designed to the same technology standards to carry the same payload for the same distance at the same cruise speed and to operate by the same set of requirements. Only by holding all these items constant could equitable comparisons be made to determine the relative worth of the two fuels in commercial transport aircraft service.

In the following two sections, results of the studies of the advanced design supersonic and subsonic transport aircraft are described.

2.4.1. Supersonic Transport Aircraft

A pictorial representation of a Mach 2.7 cruise speed LH_2-fueled vehicle is shown in Figure 5. It is designed to carry 234 passengers 4200 nmi. This design requirement will be used as a basis for the discussion which follows. It must be pointed out that the supersonic aircraft designs whose characteristics are shown here were derived in 1974 and 1975 for the Jet A and LH_2-fueled versions, respectively.[13] In the interim there has been continuing effort on Jet A-fueled concepts and many design improvements have been achieved, most of which would apply to aircraft designed for either fuel. However, no corresponding effort has been funded for development of improved LH_2-fueled SST designs, so comparisons based on the more advanced concepts cannot be drawn.

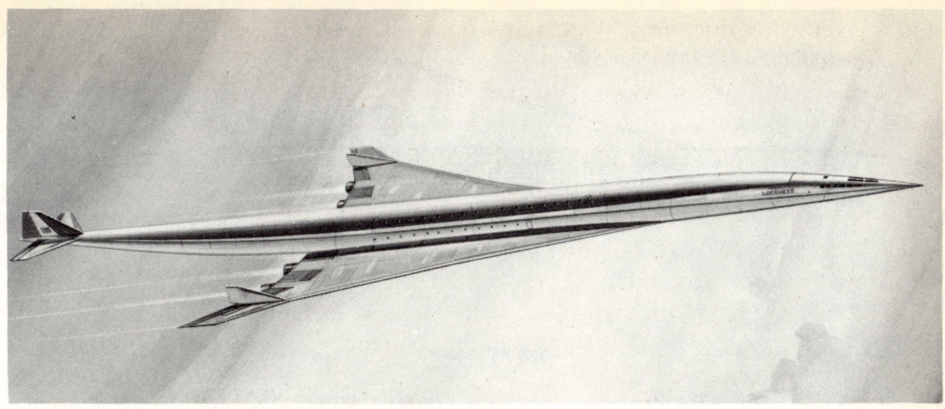

FIGURE 5.   Mach 2.7 LH₂ supersonic transport.

Some of the features of the LH$_2$-fueled M2.7 design are illustrated in the general arrangement drawing of Figure 6. The passengers are located in the central section of the fuselage directly over the wing. They are seated in a double-deck arrangement which, as shown in the inset view, Section A-A, has 3 seats on either side of a center aisle on each deck, for a total of 12 seats per row. LH$_2$ fuel is contained in tanks located both forward and aft of the passenger compartment. The double-lobe cross-section of the fuselage extends through both tank compartments.

A unique feature of this design is that there is no provision for physical access between the passenger compartment and the flight station. Although this is a departure from current practice, consultation with airline representatives concerning its acceptability produced no strong arguments requiring such access. The aircraft design using this passenger/tankage arrangement was adopted only after safety and performance evaluation of many alternate aircraft configurations. For example, designs were investigated which involved carrying a portion of the LH$_2$ fuel in the aircraft wing, similar to concepts devised for conventional Jet A fuel.

The results showed that it was inefficient to do this, even when maximum thickness ratio wings were considered, because of the large surface-to-volume ratio of the tanks. Configurations with thin, efficient wings and with fuselage cross-sectional areas large enough to contain all the fuel and the payload in a reasonable length provided superior performance.

Thus the design shown in Figure 6 was found to offer significant weight, performance, and cost advantages. Non-access between flight station and passengers is not necessarily advocated as an advantage, but neither is it regarded as a serious disadvantage in commercial transport aircraft. This point is discussed further in connection with subsonic designs.

Aside from the double-lobe fuselage and the location of hydrogen fuel tanks fore and aft of the passenger compartment, the aircraft design shown in Figure 6 is conventional and quite similar to that of its counterpart designed for Jet A fuel. The wing has the NASA-specified arrow-wing planform. The design concept of the wing structure is identical to that of the corresponding Jet A-fueled aircraft, except to account for the following differences: smaller wing area, lower wing loading, and no fuel containment. This latter point has both advantages and disadvantages; there is no need to modify an otherwise ideal load path to provide for tankage requirements; on the other hand, there is no load relief to apply at the design condition involving a 2.5g

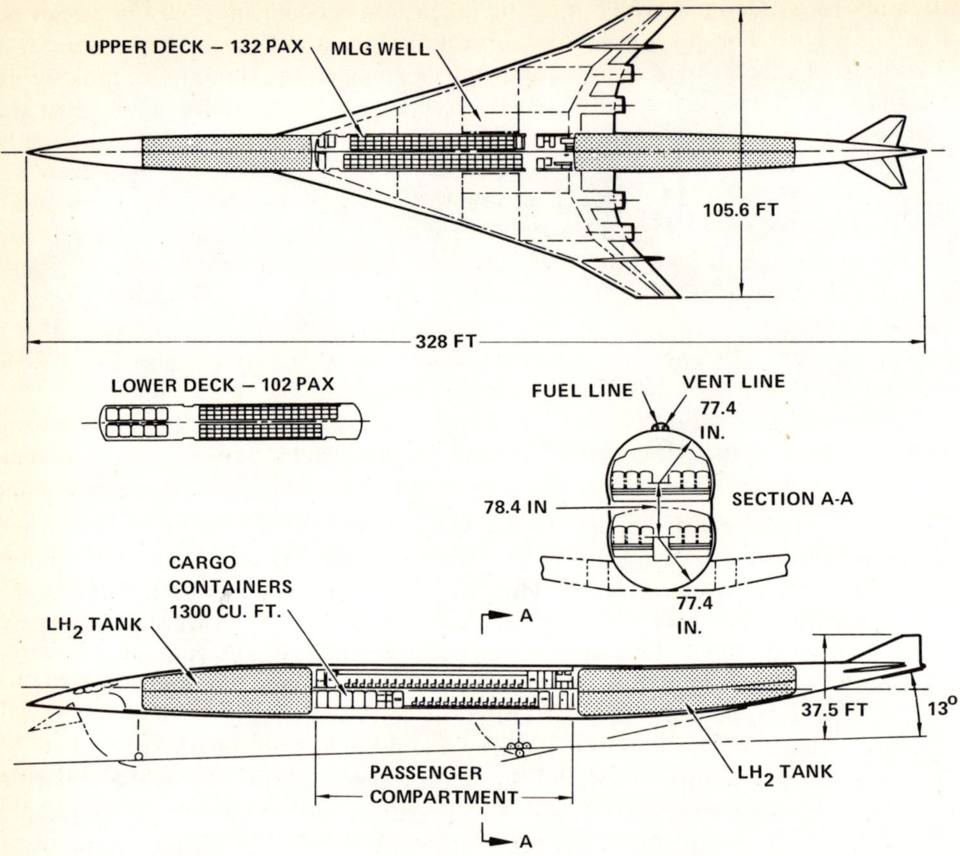

FIGURE 6.   General arrangement, LH₂ Mach 2.7 supersonic transport.

maneuver so the wing specific weight is slightly greater than would otherwise be the case.

The wing has subsonic leading edges equipped with Krueger leading edge flaps on the outboard panels. Along the trailing edge there are conventional flaps and spoilers inboard; flaperons, which function as both flaps and ailerons, are located between the inboard engine and the wing vertical surface; and there are conventional ailerons on the outboard panels for low-speed control. The fixed vertical fins located near each wing tip enhance high-speed directional stability.

The all-moving horizontal stabilizer has a geared elevator and the all-moving vertical tail has a geared rudder. Tail surfaces were sized using the same tail volume coefficient previously established for the conventionally fueled counterpart aircraft. On the LH₂ design, the physical size of the tail surfaces is smaller because the moment arm to the center of gravity of the airplane is greater.

Four turbofan engines with axisymmetric, mixed-compression inlets are mounted beneath the wing with their exhaust ducts extending just aft of the wing tailing edges. Landing gear design is conventional. The visor nose tips down to provide the crew with good visibility for takeoff and landing. A dorsal fairing above the fuselage carries fuel supply and vent lines from the vertical tail forward to the front LH₂ tank. Being above the fuselage, any fuel vapors released would rise away from the aircraft and present minimum hazard.

In hydrogen-fueled aircraft, there are many important differences, relative to conventionally fueled designs, which must be taken into account, not only in design but also in service use. For example, in the subject aircraft, a nominal tank pressure of 21 psia was assumed. This represents a compromise intended to (1) minimize tank weight and (2) provide adequate margin of positive pressure over atmospheric, allowing reasonable operating tolerances for pressure sensors and valve operation. At 21 psia, LH_2 will reach the condition of a saturated liquid at $-421.3°F$ (21.5 K). As previously discussed, with conventional fuel air is admitted to the aircraft fuel tanks to maintain approximately constant gauge pressure. The tanks vent as the aircraft ascends and admit air as the aircraft descends. With LH_2, air cannot be admitted to the fuel tanks because it would immediately freeze.

Another important difference is that when an LH_2 aircraft is being refueled or is out of service for short periods, e.g., less than a week, a vapor return line will be hooked up to capture boil-off hydrogen for reliquefaction.

Some LH_2 will be kept in the tanks at all times to maintain the system at cryogenic temperature, thus avoiding subjecting the tank structure and support system to extreme and repetitious temperature cycling and eliminating the requirement for expensive and time-consuming chill-down operations. As will be described subsequently, the fuel tanks are carefully insulated to minimize loss of hydrogen by boil-off and to prevent frost build-up on the external surfaces. That GH_2 which is vented from the aircraft tanks to maintain design pressure during refueling and out-of-service periods is recovered and reliquefied.

For extended out-of-service periods, e.g., when some type of major maintenance not related to the tank or insulation system is required on the airplane, the tanks will be defueled and purged with nitrogen but will be maintained at a pressure slightly greater than ambient to prevent air from entering. For inspection or repair of the tank structure or insulation system, after nitrogen purge the tanks will be vented to the atmosphere and can then be safely entered by maintenance personnel.

Operational procedures for LH_2-fueled aircraft are conceived as being not radically different from current practices. The equipment would be different, of course, but the manpower and the elapsed time per function should be virtually the same. During a routine fueling process, estimated to require about 30 min for a normal turn-around, cabin attendants can perform housekeeping chores, cargo can be loaded, and food service stowed. Upon completion of these services, the passengers can board, and the flight would be ready for takeoff. With properly designed equipment and scheduling of operations, there is no obvious reason a hydrogen-fueled airplane should require more time for turn-around than conventional Jet A-fueled aircraft.

2.4.1.1. Fuel System

A schematic diagram of a feasible aircraft LH_2 fuel system is illustrated in Figure 7. It consists of two pressurized and insulated tanks (each divided into two compartments), a vent system, insulated fuel supply lines to each of four turbofan engines and one auxiliary power unit, heat exchangers to transfer airframe and engine heat loads to the cryogenic fuel, fuel quantity gauging equipment, refueling and defueling systems, and an optional fuel jettison system, in addition to associated pumps, valves, seals, etc. In the following paragraphs, the function of the principal elements of the fuel system are defined. Following that, the design and physical aspects of the tanks and their thermal protection system are described.

The fuel tanks are maintained at their nominal pressure of 21 psia by an absolute pressure regulator located between the common vent line for all fuel tanks and a lightning-protected vent outlet with a flame arrestor which permits overboard discharge of gaseous boil-off at the top of the vertical tail without the hazard of flame propagation back to the fuel tanks. If the tank pressure drops below 18 psia because of exception-

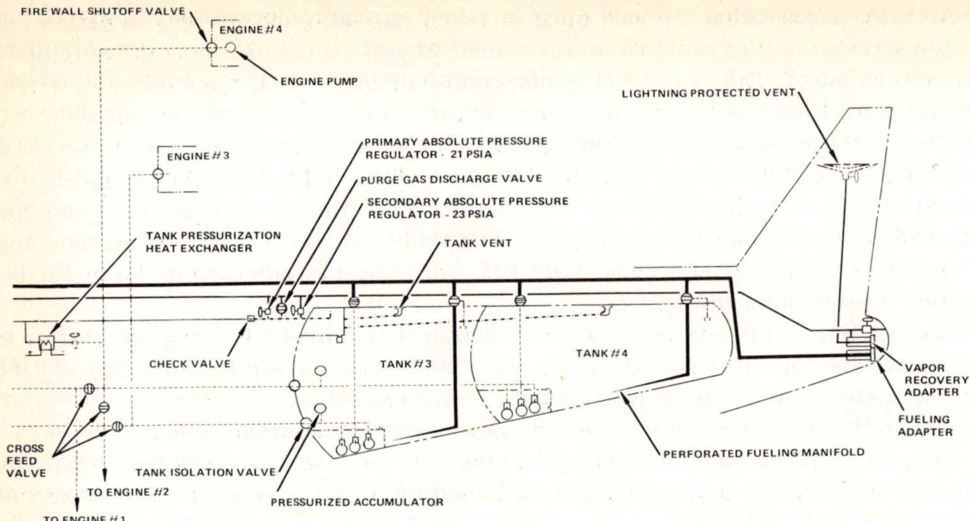

FIGURE 7.   LH$_2$ aircraft fuel system schematic.

ally high engine fuel demand, a secondary absolute pressure regulator located in the No. 4 engine feed line opens, allowing a small amount of fuel at pump discharge pressure to be vaporized before it is conveyed to the tanks through the normal vent system.

In the event tank pressure exceeds 22 psi above free stream ambient, a pressure relief valve opens to bleed off the excess pressure through the vent system flame arrestor. A tank rupture disc is also provided in the case of a dual failure of both the tank pressure regulator and the pressure relief valve. If, for any reason, the tank pressure falls below ambient outside pressure, suction relief is provided at 2 psi below ambient as an emergency measure to prevent collapse of the tanks. This condition could exist only if all of the fuel had been exhausted during a descent and if the normal vent closure did not occur or as a result of an extended ground stand-by with empty fuel tanks, again if the vent valve was not actuated properly.

Vent openings are located in the forward and aft ends of each tank. Float-operated vent valves in the opening nearest the vent box prevent fuel from flowing by gravity into the vent box. Liquid fuel which collects in the vent box is drained into the adjacent fuel tank through a float-operated drain valve.

Each fuel tank compartment is normally connected to a specific engine. However, a cross-feed system permits any one tank to supply fuel to any engine if required or, by proper sequencing of the cross-feed and refueling valves, permits transfer of fuel from one tank to another.

Three boost pumps are located in a surge box in each tank to ensure fuel availability to each engine during takeoff and to prevent fuel starvation during aircraft maneuvering at low fuel levels. The boost pumps are designed to pump boiling hydrogen and to supply it to the main engine pumps in a subcooled state by means of insulated feed lines. All airframe and engine heat loads, with the exception of the tank pressurization heat loads, are added downstream of the high-pressure engine pumps.

The auxiliary power unit (APU) can be operated on GH$_2$ from the common tank vent line, thereby minimizing boil-off losses during the considerable periods of APU operation while on the ground. If insufficient boil-off is released from the tanks, e.g., due to the presence of supercooled hydrogen just subsequent to refueling, operation of the No. 4 tank-mounted boost pump will maintain gas flow through an auxiliary heat exchanger at 18 psia to the APU.

All tanks are refueled through a pressure-fueling adapter located in the aircraft tail cone (see Figure 7). Inside each tank compartment, a portion of the fueling manifold is perforated to distribute the LH_2 uniformly along the bottom and below the surface of the nominal reserve level to minimize boil-off. A dual fuel-level control pilot valve in each tank stops the flow of fuel to that tank when it has reached its full level at approximately 96% of total volume. Integral with the float valve is a solenoid valve which permits manual or preset shut-off of the valve at any tank level and which also prevents overfilling in the event of a float valve failure.

A boil-off recovery adapter and appropriate valving are provided as part of the fueling adapter in the tail of the aircraft to permit the operator to return gaseous boil-off to ground facilities for reliquefaction instead of venting it out the normal discharge in the vertical tail. This minimizes the economic penalty resulting from loss of GH_2 during refueling and from boil-off during periods when the aircraft is out of service. It also improves safety because no gaseous hydrogen is discharged overboard.

Prior to refueling tanks that have contained air, the fueling system must be purged through the fueling adapter by an inert medium (e.g., gaseous nitrogen) to remove all oxygen, followed by GH_2 to remove all inerting gas. The purge system will utilize the purge gas discharge valve to discharge the purge gases around the pressure relief valve and overboard through the lightning-protected vent outlet.

Defueling may be accomplished through the defueling valve to the fueling adapters by operating the boost pumps with open cross-feed valves. The tanks may be defueled individually or simultaneously.

Because of its low fuel weight, the LH_2-fueled aircraft will not require a jettison system to meet the climb requirements of Federal Air Regulation 25. However, some situations can be postulated in which a jettison system might be desireable. The fuel system illustrated in Figure 7 permits jettison to be accomplished. It operates in the same manner as the defueling system, with the fuel being expelled out the tail cone of the fuselage.

Capacitance gauges are shown as a representative system for measuring fuel volumes. The units would be calibrated to indicate fuel quantity in pounds at the fuel management panel.

As a design objective, fuel system components such as pumps and valves would be designed for quick removal and field replacement in a manner commensurate with present commercial operation. Provisions would also be made so that in the event of system failure, back-up for critical dispatch items would be available as in current practice.

For reasons of flight safety, fuel system components and their arrangement and location in the aircraft must be considered in terms of malfunction and leak detection, isolation, inerting and/or purging requirements, and fire containment. Safety criteria and acceptable design practices must be established by analysis and experimental development.

2.4.1.2. Fuel Containment System

The design and development of the fuel tanks and their thermal protection systems to contain the cryogenic LH_2 in a satisfactory manner is regarded as one of the crucial technical challenges confronting use of LH_2 in operational aircraft.

Materials used for tank construction must be resistant to hydrogen embrittlement; impermeable (or capable of being sealed) to GH_2; and, depending on the insulation arrangement, capable of retaining satisfactory ductility and fracture resistance at cryogenic temperatures. In addition, the materials must be amenable to repair and maintenance.

The process of selecting a structural concept for the tanks must involve consideration of special problems related to the nature of LH_2 as well as the fundamental problem of design for light weight while maintaining adequate structural integrity. For example, the influence of differential thermal expansion on aircraft structure is dependent on the insulation system employed for the fuel tanks. If cryogenic insulation can be applied to the inside of the tanks, the tank walls remain at near-ambient temperature; thus, differential thermal expansion, relative to the warm aircraft skin and primary structure, is minimized. Similarly, the problems of attachment and support for the tanks are simplified. The difficulty is that the insulation system, being constantly exposed to hydrogen, must be impermeable so GH_2 cannot diffuse to the tank wall, thereby raising the thermal conductivity of the insulation to that of hydrogen and crippling its effectiveness. There is no known plastic insulant which is impervious to GH_2. On the other hand, if the cryogenic insulation is applied to the exterior surface of the fuel tanks, the tank itself will significantly contract and expand as LH_2 at $-423°F$ is introduced and used. Attachment problems for the structural support system are therefore severe, not only because of the dimensional changes which must be accounted for, but also because of the potential for thermal leaks. External insulation is more susceptible to mechanical damage. In addition, it must be impervious to air, a less rigorous requirement than that faced by the internal insulation but still vital to prevent cryopumping.

The necessity of being able to inspect the tanks and maintain them in accordance with airline standards is an aspect which cannot be overlooked. Finally, in those designs where the fuel tanks are within the aircraft framework and where it is therefore possible for leaking gaseous hydrogen to collect in a confined space, provision must be made for a purge system using either an inert gas or copious quantities of air.

The candidate design possibilities for fuel containment systems consist of combinations of nonintegral and integral tankage, together with both internal and external insulation. Nonintegral tanks are those which are designed to take only loads associated with containment of the fuel, i.e., pressurization and fuel dynamic loads, plus thermal stresses. They are supported within conventional fuselage skin/stringer/frame structure. On the other hand, integral tanks are an integral part of the basic aircaft structure; therefore, in addition to the above loads they must be capable of withstanding all the usual fuselage axial, bending, and shear stresses resulting from the critical aircraft loading conditions.

Figures 8A and 8B are illustrations of the nonintegral and integral concepts with external insulation. The thermal protection system in each case would consist of a layer of closed cell plastic foam for cryogenic insulation, a wrap of aluminized Kapton to serve as a vapor barrier film to prevent cryopumping of air in the event the cryogenic insulation develops cracks during service. A layer of fiberglass mat sealed with a coating of polyimide resin would be provided for high-temperature insulation for the integral design.

The nonintegral tank is mounted in the fuselage structure using a four-point support as illustrated in Figure 8A. This mounting system is similar to that used in mounting aircraft engines. It provides for thermal expansion and contraction and prevents fuselage deflections from stressing the tank. The directions in which loads can be resisted are indicated at each of the four points.

One of the basic problems of the nonintegral tank concept is the difficulty of inspecting the insulation system. Since it is within the fuselage structure, the fuselage must be taken apart and the tank removed to permit examination of the external surface of the thermal protection system. Fuselage break points and a removal rail are shown in the figure.

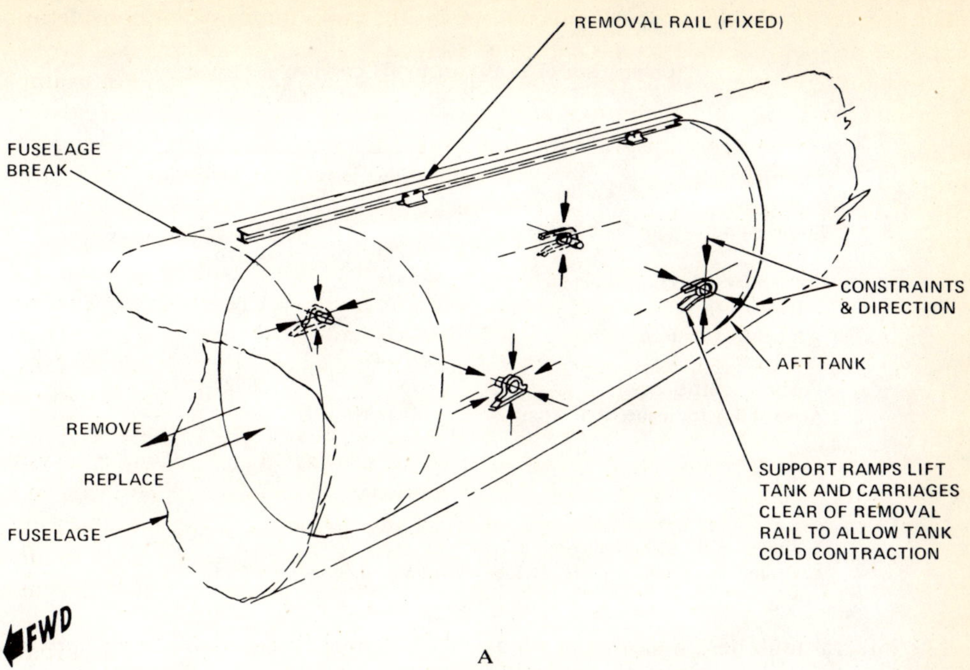

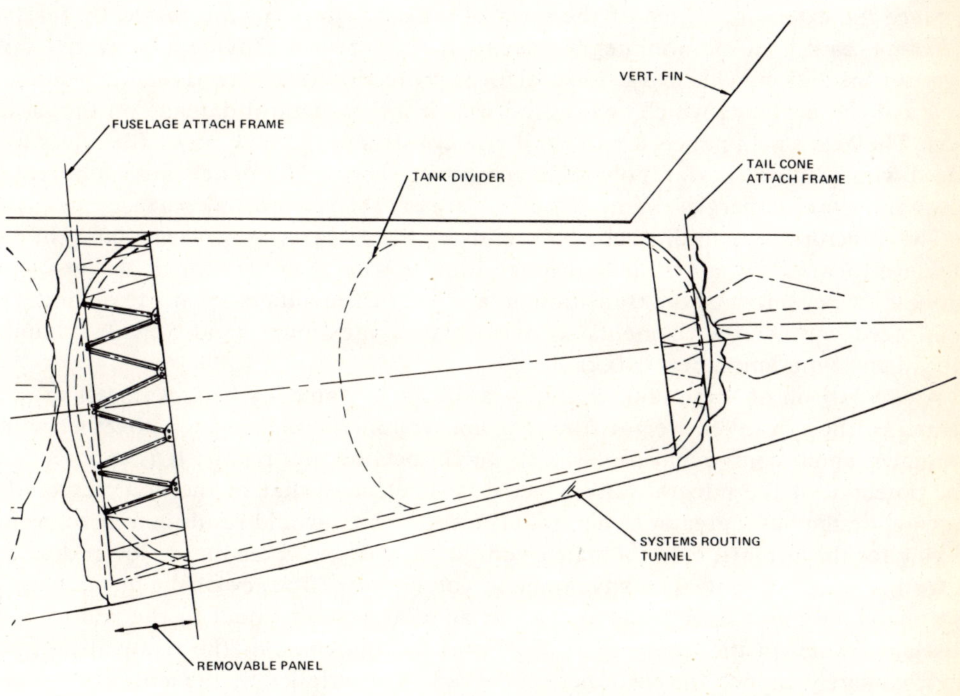

FIGURE 8.   Tank Structural Concepts. (A) Nonintegral. (B) Integral.

TABLE 4

Comparison of Tank Structural Concepts

	Tank concept	
	Nonintegral	Integral
Fuel weight fraction (lb/lb of LH_2)		
Tank[a]	0.113	0.196
Thermal protection system	0.079	0.06
Heat shield	0	0.06
Fuselage structure	0.152	0
Total	0.344	0.316
Volumetric efficiency[b]	0.855	0.927
Accessibility for inspection/repair	Requires removal of tank from aircraft	Remove heat shield

[a] Based on 40,000 psi allowable stress.
[b] Volume available for LH_2 divided by volume of fuselage section.

The integral tank design shown in Figure 8B is similar to the nonintegral, but with important differences. The basic differences are that the integral tank has longitudinal stiffeners, it has a heat shield added to the thermal protection system, and it is attached to the fuselage sections forward and aft of the tank by a truss framework.

Since the external surface of the integral tank design is not protected by fuselage structure, as it is in the nonintegral case, it is necessary to provide a protective layer over the insulation. This extra shield affords protection from aerodynamic heating as well as from air loads which could otherwise inflict structural damage on the insulation. The heat shield panels are of sandwich construction, made up of fiberglass filler faced with graphite/kevlar/polyimide composite sheets. The panels are supported by low-conductance fiberglass standoffs which are fastened to the tank surface.

The structural transition from both ends of the integral tank to the conventional fuselage forward and aft is made through tubular truss members which are positioned around the periphery of the transition area. The truss members are made of fiberglass reinforced with boron filaments so as to afford maximum rigidity with minimum weight and a minimum heat leak rate.

A comparison of these tank concepts is shown in Table 4 where estimates are presented of their installed weight fractions and volumetric efficiencies, together with a comment about component accessibility for inspection and repair. It is apparent that the potential of the integral tank concept is superior to that of the nonintegral. The integral design has a greater structural efficiency which would result in a direct weight saving for the aircraft. It has a higher volumetric efficiency, leading to lower drag and a further weight saving. The advantage in volumetric efficiency of the integral design is apparent when it is considered that an annular volume equal to the width of the fuselage frames in the nonintegral case, plus the thickness of the removal rail, plus approximately an inch for clearance on the radius, multiplied by the length of the tank less its ends, is lost to the nonintegral concept.

In addition, the tank/fuselage structure, as well as the insulation, is more readily accessible for inspection and repair. Tank structure of both design concepts can be inspected from inside the tanks; however, any repairs required by the nonintegral type would necessitate removal from the airplane. The aircraft fuselage would have to be

TABLE 5

Integral Tank Weight Breakdown

Tank component (lb)	Fatigue stress level	
	40,000 psi	30,000 psi
Shell	8,700	11,580
Ends	1,350	1,800
Frames	1,970	2,120
Baffles/bulkheads	810	810
Crosstie	900	1,078
Transition trusses	2,500	2,500
Crack stoppers	240	320
Contingency (10%)	1,660	2,020
Total (both tanks)	18,130	22,228
Fuel weight (lb)	92,500	92,500
Tank weight fraction $\dfrac{W_{tank}}{W_{LH2}}$	0.196	0.2403

TABLE 6

LH₂ Tank Sizing Allowances

	Percent
Tank chill-down contraction	0.90
Structure allowance	0.52
Equipment allowance	0.08
Fluid expansion (15 to 21 psia)	1.70
Ullage after expansion	0.30
Unuseable	0.30
Pressurant gas	1.77
Vented boil-off	1.63
Total allowance	7.20
Effective density of tanked LH₂	4.118 lb/ft³

designed with break-joints so that the aircraft could be pulled apart and the tanks completely removed, even for periodic inspection of the tank insulation. Any imperfections in the insulation of the integral design would be marked by obvious displays of frost since the thermal protection system of that tank concept is the external surface of the aircraft.

Local repairs to the integral tank can be effected by simply removing the thermal protection system in the area requiring attention.

For these reasons, the integral tank concept was selected for incorporation in the final aircraft design. It should be recognized that the specific fabrication concept shown, i.e., the integrally stiffened shell design, is not necessarily advocated; it is merely presented as an example. There are many alternatives which should be explored. For example, a brazed aluminum honeycomb sandwich structure is a possibility, as is a weld-bonded skin stringer construction. Either of these would require more extensive development of fabrication technology than would the concept shown.

Subsequent to the initial comparative analysis of tank structural concepts, the results of which were shown in Table 4, the allowable stress level for the tank structure was reduced from 40,000 psi to 30,000 psi. This change was based on consideration of fatigue characteristics commensurate with aircraft design life of 50,000 hr in commercial airline service. Table 5 shows a breakdown of weights of the tank components, total weight, and tank weight fractions relative to the weight of fuel contained for both stress levels. The weights listed are for the total of both the forward and aft tanks. A contingency of 10% has been included in both columns to account for unknowns and uncertainties. The tank weight fraction increased from 0.196 to 0.240 due to the decrease in allowable stress.

Allowances for sizing the tanks are listed in Table 6. The volume required is based on the "as built" (warm) condition and includes the other allowances listed. The fluid expansion to 21 psia pressure is assumed to occur after filling from a ground storage facility at an equilibrium temperature corresponding to 15 psia. The effective density shown was used to calculate the tank volume required for the total amount of fuel which must be contained in the aircraft to fly the design mission, including reserves and boil-off.

TABLE 7

Plastic Foam Insulation Materials

Material	Density (lb/ft³)	Thermal conduc- tivity (Btu/hr ft °F)
Polymethacrylimide — rigid, closed cell (Rohacell® 41S)	2.2	0.018 (at 43°F)
Polyurethane — rigid, open cell	2.0	0.021 (at 39°F)
Polyurethane + 10% chopped — glass fibers, rigid, closed cell (A.D. Little Co.)	4.0	0.012 (at −20°F)
Polyvinylchloride — rigid, closed cell (Klegecell H917)	3.1	0.0087 (at −110°F)

With the basic design concept of the tank established, thereby defining the operating conditions and requirements for the thermal protection system, a more detailed examination of the design of the insulation system could be made. In addition to the fundamental requirement that the insulation system control hydrogen boil-off to acceptable levels, it must also prevent frost build-up on the aircraft.

Further, an acceptable insulation system must have mechanical properties which will enable it to operate satisfactorily in a commercial airline environment for the required life span with a minimum of maintenance. To summarize these requirements, a commercially acceptable thermal protection system would use insulation materials that are (1) impervious to air so they do not require purging to prevent cryopumping, (2) not susceptible to aging or cracking under repeated thermal or mechanical stresses, (3) able to withstand the exterior temperatures associated with supersonic flight, and (4) capable of being repaired or replaced readily.

In addition, because of the severe consequences of even a tiny leak, designs which are dependent on maintaining a high vacuum are not considered acceptable, thus eliminating the conventional multilayer and perlite systems.

A promising new concept is under development at Lockheed Missiles and Space Company which involves use of borosilicate microspheres in a partially evacuated annulus. Early test results show acceptable conductivities can be obtained with partial pressure much less severe than that used with either perlite or multilayer insulation, with an annulus thickness of only 1 or 2 in., and with a competitive system weight. With continued development, systems such as this should offer significant advantage over materials and designs currently available.

The rigid foam insulants listed in Table 7 were considered the most promising of the currently available candidates for the subject aircraft applications. The polyvinylchloride, reinforced polyurethane, and polymethacrylimide closed-cell foams were preferred. As previously stated, however, none of these is impervious to GH_2, so consideration of the internal insulation system must await development of a suitable material. Accordingly, attention was focused on external insulation systems. The characteristics of Rohacell® 41S, the polymethacrylimide foam with fire retardant additives, were arbitrarily selected to represent this general category of insulation and were used to determine the thickness of material which should be applied to the outside of the hydrogen tanks of the subject aircraft to provide acceptable thermal protection for the least weight and cost.

The trade-off of insulation thickness vs. weight of insulation, hydrogen boil-off, and heat shield is shown in Figure 9 in cumulative fashion. As noted on the figure, for this calculation the vehicle was not resized to permit carrying the fuel load required to perform the design mission as the weight of the thermal protection system and boil-off losses varied. An insulation thickness of approximately 3 in. is shown to represent the minimum weight situation.

The effect of insulation thickness on cost of operating the aircraft is shown in Figure 10. In this presentation, the aircraft has been resized to fly the design range as insulation thickness is varied. The lowest curve on the figure is the difference in aircraft cost amortization over a 50,000-hr life cycle which results from aircraft size and weight changes due to carrying insulation of various thicknesses, including consideration of corresponding losses of hydrogen due to boil-off. The cost for block fuel used during

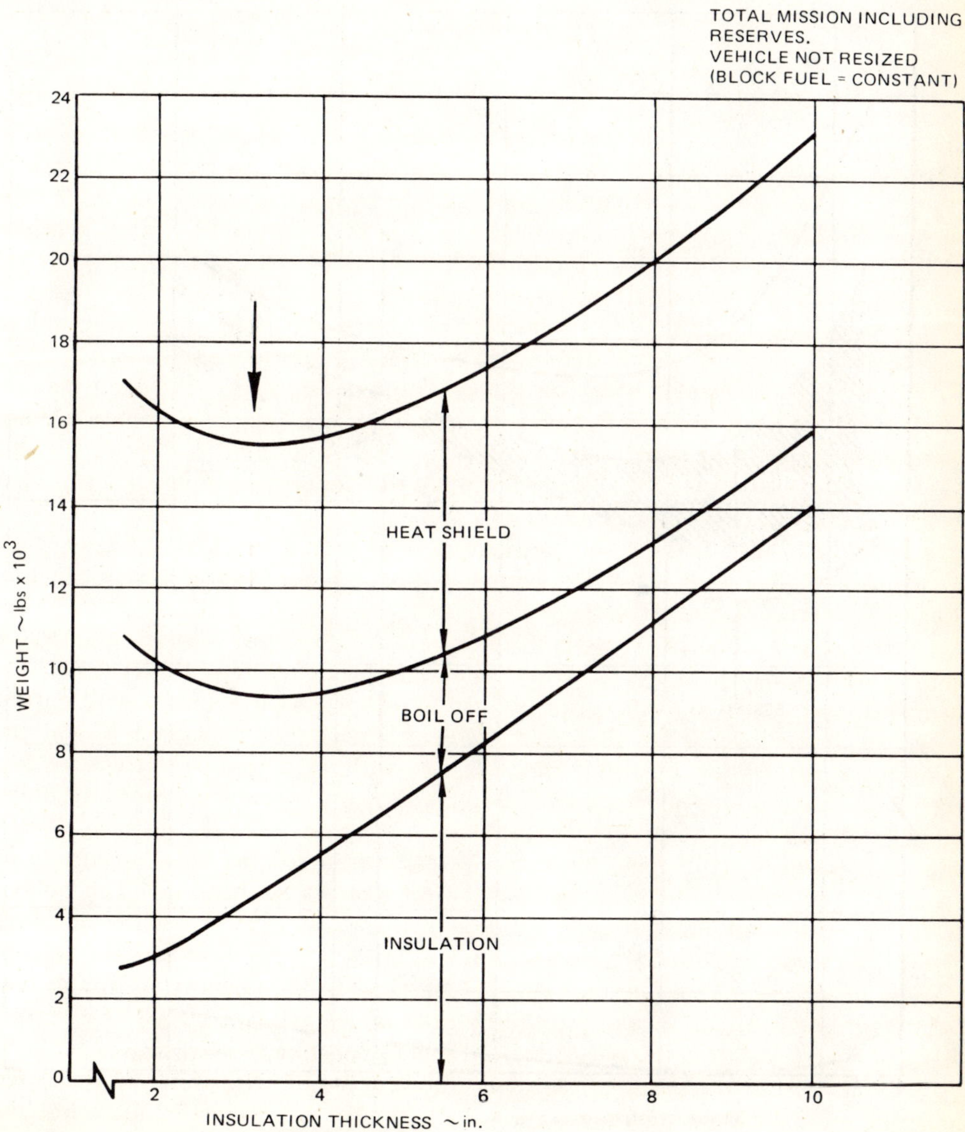

FIGURE 9.   Insulation thickness vs. weight for Mach 2.7 transport.

a flight and of hydrogen boiled off both during flight and on the ground is presented in terms of cost per day as a function of insulation thickness. The minimum point in the top line, the cumulative effect of all factors, occurs at an insulation thickness of about 5 in. These results were obtained on the basis of no recovery of boiled-off hydrogen on the ground, i.e., as if vent gases were simply allowed to escape. In comparison, the mimimum point in the second curve from the top, which includes in-flight boil-off but in effect assumes 100% recovery of ground boil-off, occurs at about 4.5 in. of insulation thickness. It also shows that recovery of ground boil-off hydrogen can make a difference of about $1000 per day based on the cost specified for LH_2 at the time of the study ($3 per million Btu). A portion of the GH_2 boiled off in flight is

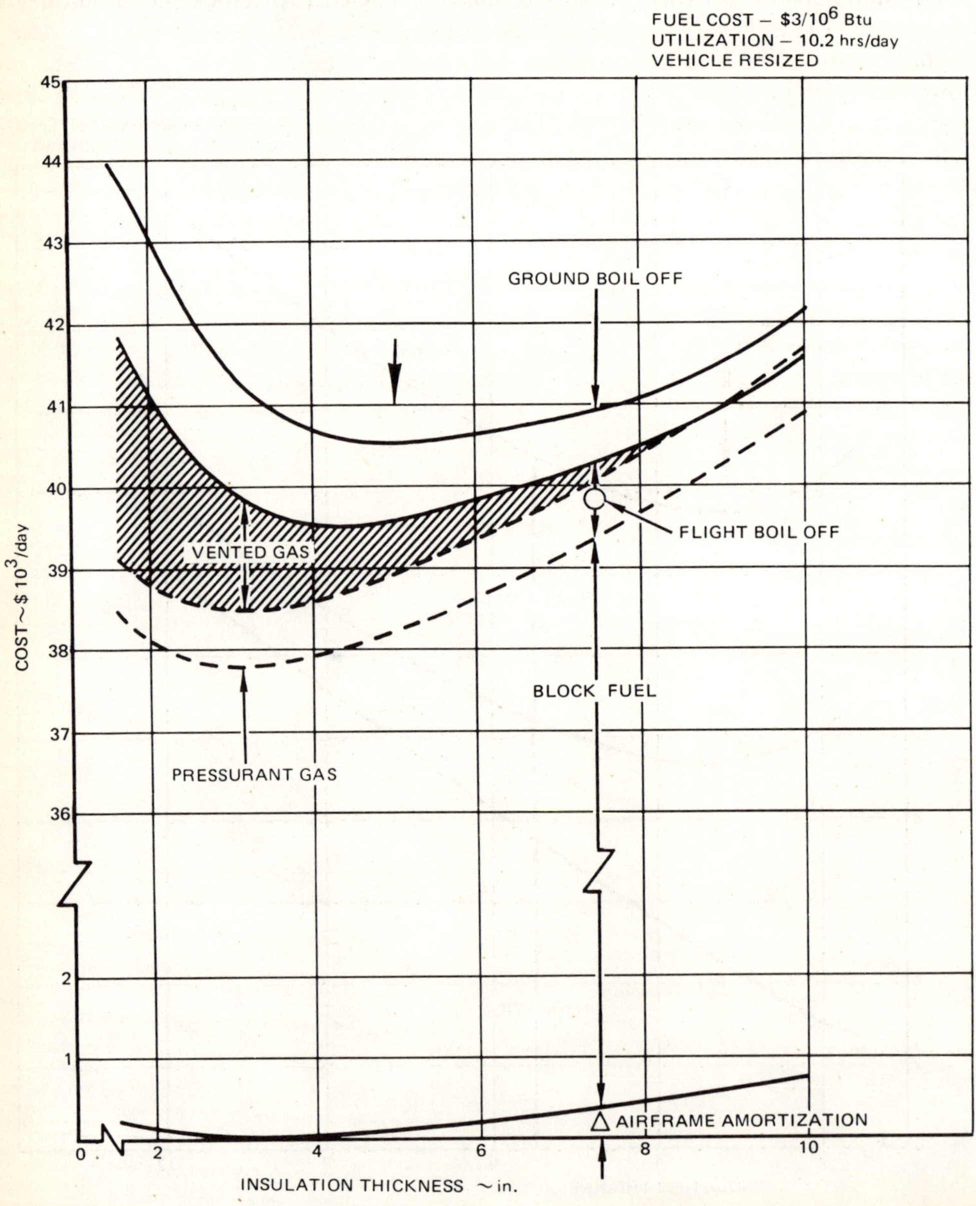

FIGURE 10.   Insulation thickness vs. cost for Mach 2.7 transport.

required to maintain design pressure in the tanks. The amount boiled off in excess of this requirement, and therefore presumed to be vented overboard, is indicated by the shaded portion of in-flight boil-off. Based on these results, a nominal thickness of 5 in. of foam insulation was selected to serve as a basis for performance and cost evaluations of the Mach 2.7 SCV.

2.4.1.3. Engine Characteristics

Some of the effects which use of hydrogen as fuel might have on aircraft engine design and performance were reviewed qualitatively, earlier. Specific characteristics of an advanced design, duct-burning turbofan engine, which were derived for use in the aircraft parametric design study are presented in the Figures 11 to 17 and Table 8.

A cycle optimization study yielded values of turbine inlet temperature (TIT), fan pressure ratio (FPR), and overall pressure ratio (OPR) which combined to produce the most favorable performance for an LH_2 fueled turbofan engine. The results are shown in Figure 11 where the effect on specific fuel consumption (SFC)resulting from varying each of these engine parameters one at a time is plotted for three different flight conditions. Based on these results and component performances and efficiencies selected to represent 1985 state-of-the-art advanced engine technology, the cycle characteristics shown in Table 8 were derived for the subject engine. The bypass ratio for this design is 4.41.

Typical calculated engine performance data are shown in Figures 12 to 15. Figure 12 presents thrust at takeoff power setting at sea level and three other altitudes as a function of speed. As noted on the figure, the thrust was calculated using the maximum duct temperature of 2460°R and for a standard day temperature plus 27°F, viz., 86°F.

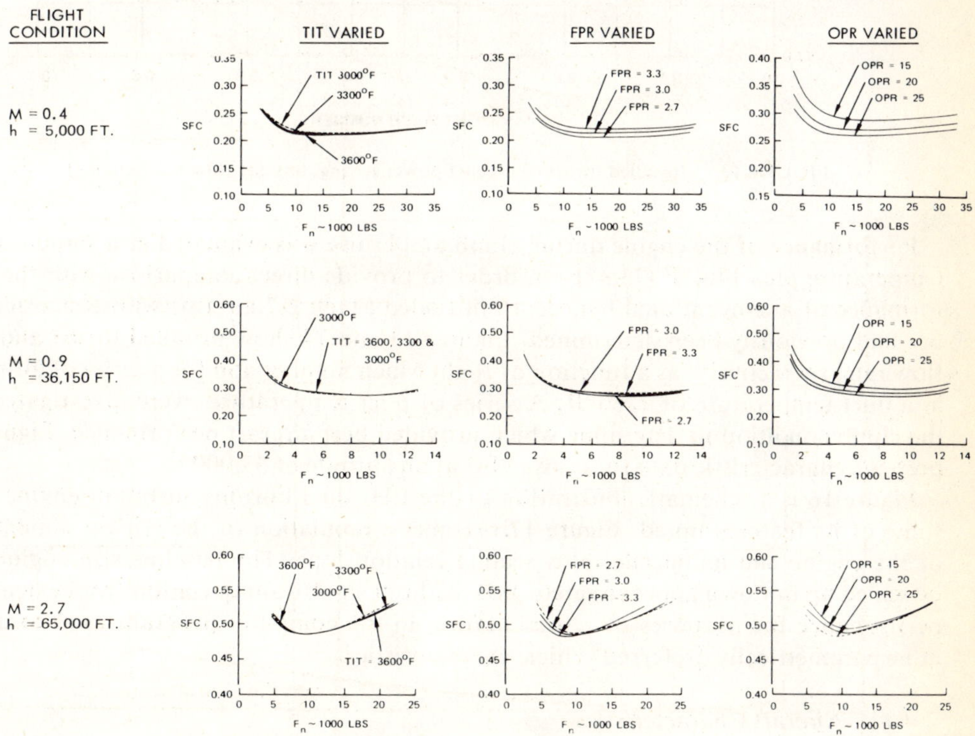

FIGURE 11.   Engine parameter selection: LH_2 duct-burning turbofan.

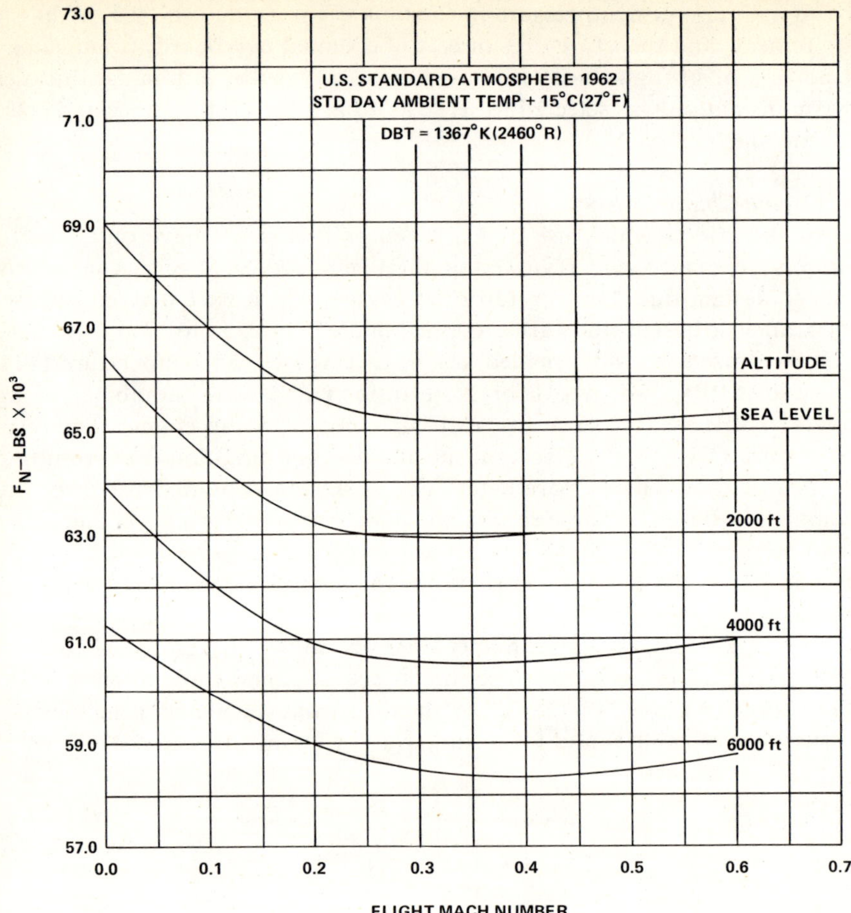

FIGURE 12.   Installed thrust — takeoff power setting, hot day (Mach 2.7 engine).

Performance of the engine during climb and cruise was evaluated at a standard day temperature plus 14.4°F (73.4°F) in order to provide direct comparison with the performance of a conventional hydrocarbon fueled Mach 2.7 aircraft whose characteristics had previously been determined. Figures 13 and 14 show installed thrust and fuel flow rate, respectively, as a function of flight Mach number and for a series of altitudes at a duct temperature of 2260°R. A series of duct temperatures were investigated for the climb condition to determine which provided best aircraft performance. Figure 15 presents characteristic data for cruise SFC at an altitude of 65,000 ft.

Figure 16 is a schematic illustration of the LH_2 duct-burning turbofan engine with some of its features noted. Figure 17 presents a tabulation of the critical dimensions of the engine and its nacelle, plus scaling relationships. The baseline size engine can be scaled up or down approximately 30% without significant deviation from calculated performance for purposes of representation in the computer program used to determine parametrically preferred vehicle characteristics.

2.4.1.4. Aircraft Characteristics

A parametric design study of supersonic transport aircraft assessed the effect of variables which included wing loading, wing thickness ratio, body cross-sectional area, thrust-to-weight ratio, and type of engine. The best of a very large number of theoret-

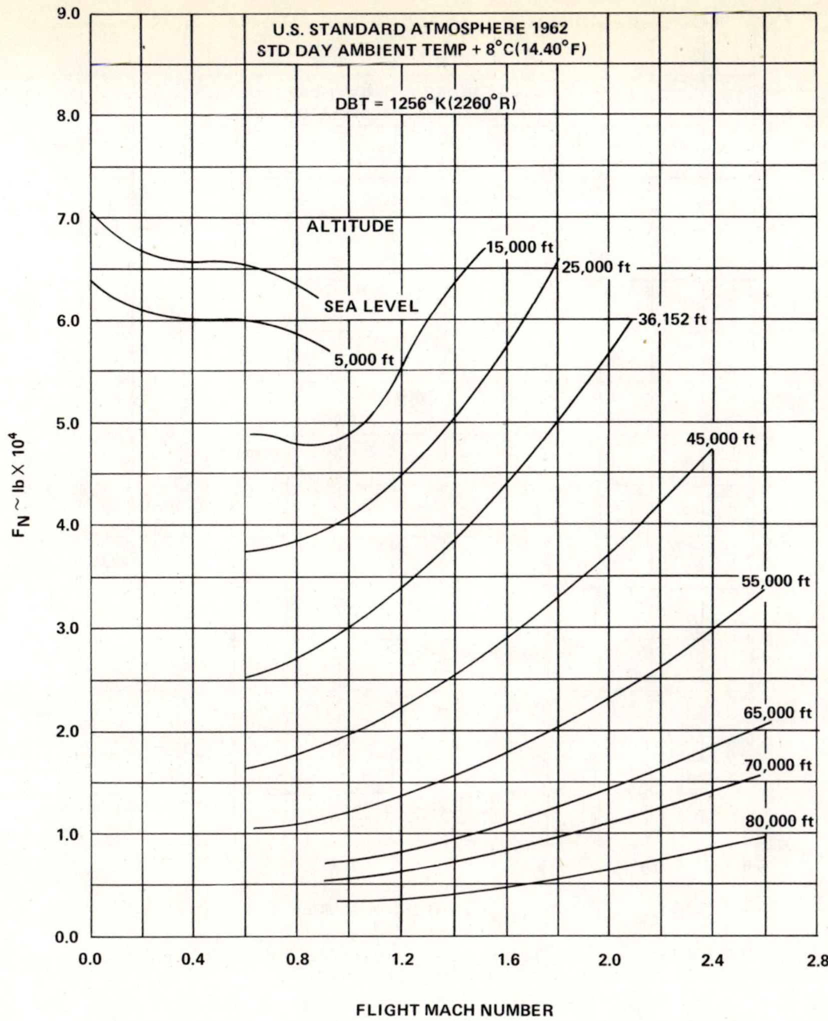

FIGURE 13.   Installed thrust — climb setting, hot day (Mach 2.7 engine).

ical combinations of these variables was selected on the basis of aircraft performance, cost, noise, energy utilization, and considerations of general feasibility and practicability for comparison with an equivalent aircraft designed to use conventional hydrocarbon (Jet A) fuel.

The general configuration and interior arrangement of the selected LH_2 fueled Mach 2.7 design were shown in Figures 5 and 6, respectively. Table 9 presents a summary of significant characteristics of both the LH_2 aircraft and its Jet A fueled counterpart. Both aircraft are designed to the same ground rules and will carry the same payload for the same distance at the same cruise speed. The designs are both based on technologies assumed to be available for use by about 1985. As seen in the table, LH_2-fueled aircraft offers significant advantages over its Jet A counterpart. For example, the conventionally fueled (Jet A) aircraft weighs almost twice as much at takeoff. This is an important advantage for the LH_2 design because it helps to minimize airline operating costs. Wheels, tires, and brakes are all sized as functions of gross weight. They are among the most significant maintenance cost items. Low gross weight also minimizes ground handling problems and cost of equipment. In addition, low gross weight means

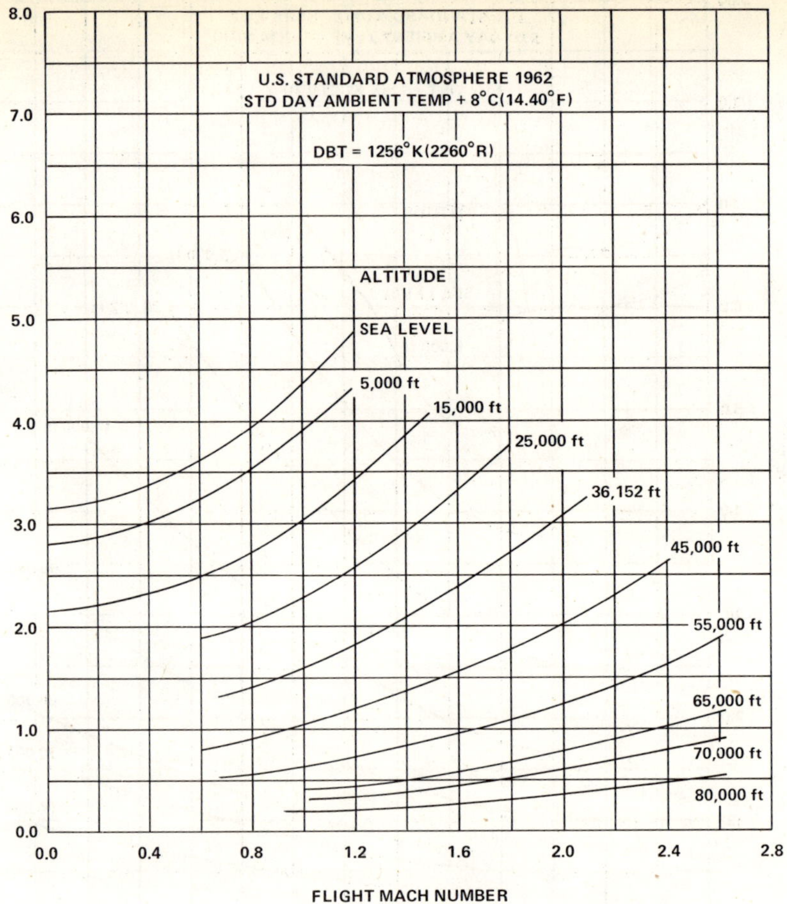

FIGURE 14.   Installed fuel flow — climb setting, hot day (Mach 2.7 engine).

smaller engines since engines basically are sized to provide the thrust/weight ratio needed to meet takeoff field length requirements, modified as needed also to meet noise limitations. Smaller engines mean lower initial cost as well as lower maintenance costs.

Operating empty weight of the LH_2 design is only 77% that of the Jet A vehicle. This is in spite of the following weight ienalties assessed against the LH_2 design to account for recognized structural and equipment differences, in addition to those mentioned earlier relative to wing design:

1. Body — Added 6% of Jet A body weight for double-deck passenger cabin and two extra pressure builkheads (the weight increment, ΔWt, is approximately 2400 lb)
2. Landing gear — 180-in. extended strut length of main leading gear in lieu of 160 in. to provide for an adequate scrape angle with a longer body ($\Delta Wt \simeq 1100$ lb)
3. Engine — LH_2 engine weight/thrust (SLS) = 0.13751 lb/lb in lieu of 0.142859 lb/lb for the Jet A design
4. Fuel system — Added 80% to weight of comparable Jet A fuel system for insulation and/or vacuum tubing around fuel lines ($\Delta Wt \simeq 2000$ lb)
5. Integral LH_2 tanks — Added $0.0958 \times$ weight of LH_2 fuel (WLH_2) for tank ends and support structure ($\Delta Wt \simeq 9645$ lb)

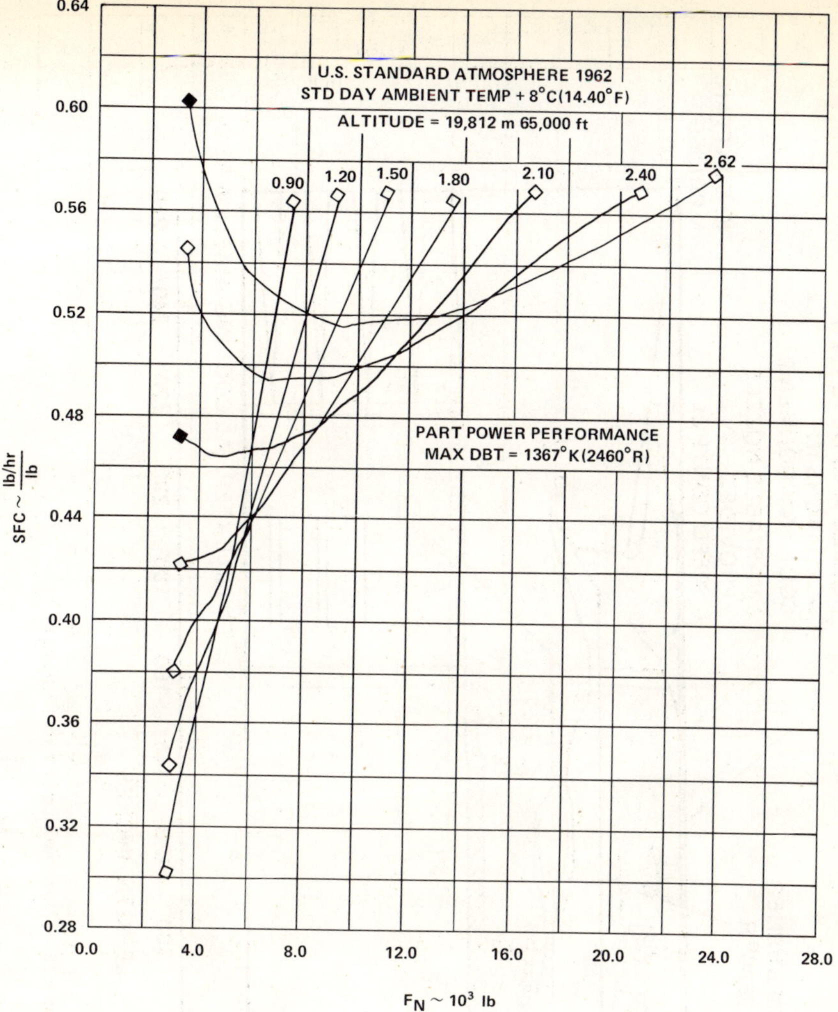

FIGURE 15.   Installed performance — cruise setting, hot day (Mach 2.7 engine).

6.   Tank heat shield and insulation — Added $0.1349 \times (WLH_2)$ ($\Delta Wt \simeq 13,531$ lb)
7.   Unusable fuel and boil-off (percent of usable fuel weight, WLH_2) — 4.10% WLH_2
     in lieu of 0.89% W Jet A

The 77% advantage in OEW represents a 32,700-lb reduction of empty weight which need not be either manufactured (at an average cost of about \$186/lb for a supersonic transport aircraft) or lifted and accelerated to cruise conditions on every flight for the life of the aircraft. These results lead to operating economies for the airlines.

Dimensionally, the LH_2-fueled aircraft has a much smaller wing, both in area and in span; however, its fuselage is larger. This is the result of carrying the hydrogen fuel in the fuselage rather than in the wing as is the case with conventionally fueled aircraft. As shown in Table 9, LH_2 has a tanked density, including allowances for ullage, cryogenic insulation, access space, etc., which is almost equivalent to the density of conventional passenger accommodations, i.e., just over 4 lb/ft.[3] The result is that LH_2-fueled aircraft are volume limited rather than weight limited. The challenge in design is to develop the most efficient packaging concept for fuel tanks, cargo, and passengers. In addition, as noted earlier, it is important to minimize the surface to volume

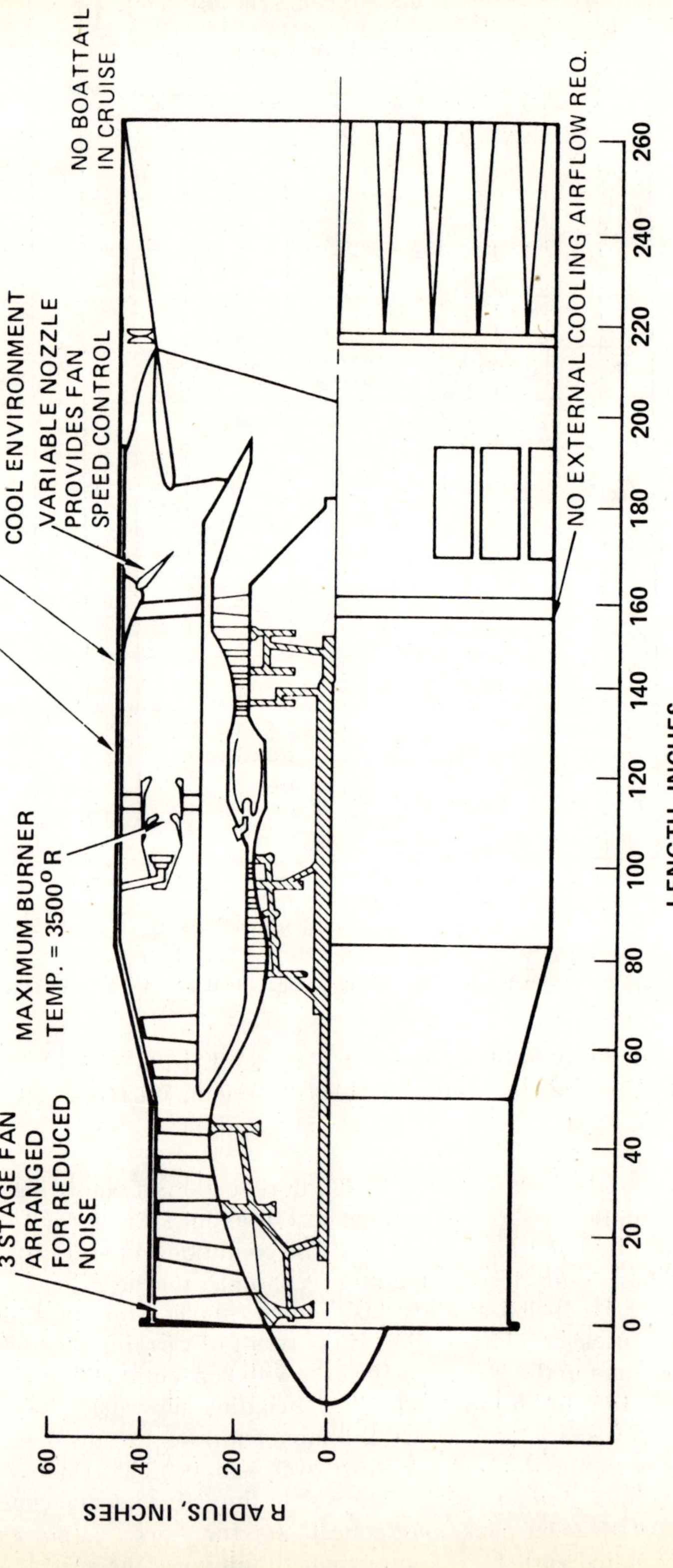

FIGURE 16.   Duct-burning turbofan schematic.

PARAMETER	REFERENCE VALUE
CRUISE MACH NO	2.7
F_N SLS MAX	84,500 lb
A_C	34.2 ft^2
D_{COMP}	81.47 in.
D_{MAX}	102.5 in.
D_{NOZZLE}	102.5 in.
L_{ENG}	267 in.
WEIGHT*	11,600 lb

* INCLUDES REVERSER AND SUPPRESSOR

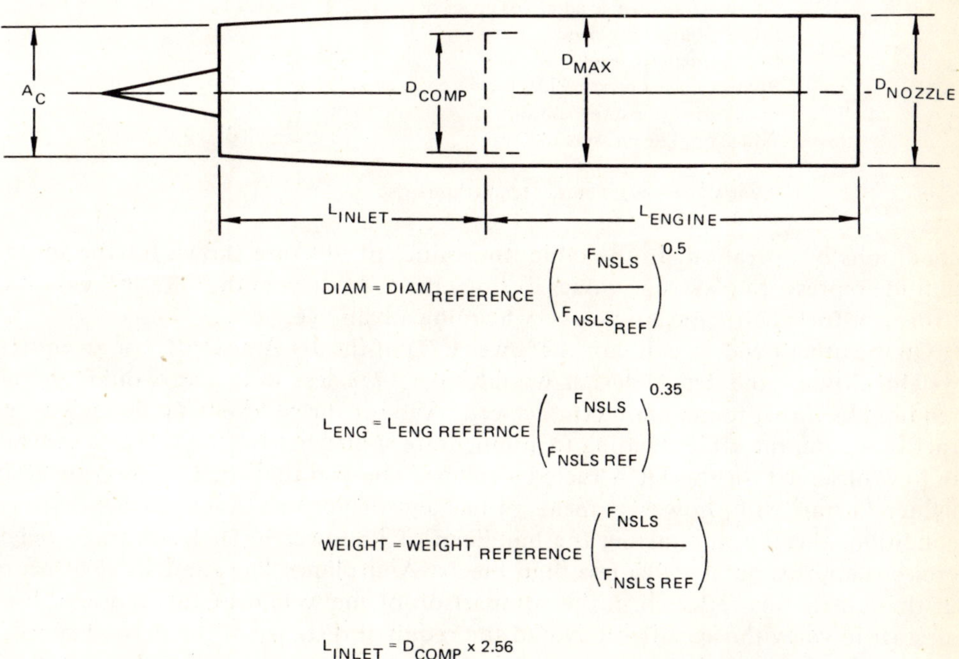

$$DIAM = DIAM_{REFERENCE} \left(\frac{F_{NSLS}}{F_{NSLS_{REF}}} \right)^{0.5}$$

$$L_{ENG} = L_{ENG\ REFERNCE} \left(\frac{F_{NSLS}}{F_{NSLS\ REF}} \right)^{0.35}$$

$$WEIGHT = WEIGHT_{REFERENCE} \left(\frac{F_{NSLS}}{F_{NSLS\ REF}} \right)$$

$$L_{INLET} = D_{COMP} \times 2.56$$

FIGURE 17.   Duct-burning turbofan nacelle reference dimensions and scaling data.

ratio of the tanks to reduce insulation weight and hydrogen boil-off. These requirements are best met by packaging the fuel in the full cross-section of the fuselage. This,
of course, results in a longer, larger diameter fuselage than for a Jet A fueled aircraft
designed for an equivalent mission.

The combination of small wing area and large fuselage which characterizes LH$_2$
fueled aircraft results in a less efficient aerodynamic configuration, indicated by the
17% penalty in lift-to-drag (L/D) ratio shown in Table 9. This parameter and specific
fuel consumption (SFC) are the operating characteristics which account for the principle differences between the two airplane designs.

SFC is a measure of the energy content of a fuel and of the effectiveness of an
aircraft engine in converting that energy to useful thrust. It is measured in pounds per
hour of fuel flow rate divided by thrust output achieved by the engine during various

TABLE 8

**LH_2 Duct Burning Turbofan Cycle Characteristics (SLS, Standard
Day, Uninstalled)**

Design cruise Mach no.	2.7
Engine type	Duct burning turbofan
Corrected airflow $W\sqrt{\theta}/\delta$	1026 lb/sec
Fan pressure ratio	3.0
Compressor pressure ratio	8.33
Overall pressure ratio	25.0
Nozzle velocity coefficient (duct)	0.981
Nozzle velocity coefficient (primary)	0.981
Maximum turbine inlet temperature	3460°R
Maximum duct burning temperature	2460°R
Fuel heating value	51590 Btu/lb
Peak fan polytropic efficiency	90.0
Peak compressor polytropic efficiency	91.5
High pressure turbine adiabatic efficiency	92.0
Low pressure turbine adiabatic efficiency	91.0
Primary burner efficiency	1.00
Duct burner efficiency	a
Primary burner pressure loss ratio	0.060
Duct burner pressure loss ratio	a
Primary nozzle pressure loss ratio	0.005

a Variable based on burner temperature rise.

conditions of operation. In the table, the values of SFC are shown for the speed and altitude representing average values during cruise. As seen in the "Ratio" column, the hydrogen-fueled airplane has an overwhelming advantage.

On the other hand, because of the lower L/D of the Jet A aircraft, for an equivalent weight airplane the Jet A design would have 17% less drag and would, therefore, require less thrust to maintain cruise speed. A lower thrust level would mean a smaller fuel flow rate, the effect tending to compensate somewhat for the severe disadvantage in SFC observed for the Jet A fueled airplane. The fact that the LH_2 aircraft design is lighter to start with, however, means it has less weight to lift and accelerate to cruise condition, thereby minimizing the handicap of its lower L/D. Its average weight in cruise is approximately 50% less than the Jet A airplane. The cumulative effect of all of these factors is reflected in the comparison of the weight of fuel required by each aircraft to carry the specified payload the required distance at the desired speed. The Jet A aircraft must take off with almost four times the weight of fuel required by the LH_2 airplane for the design mission.

Both of the supersonic transport aircraft compared in the table were designed to have an equivalent approach speed of 158 kn. This requirement established the wing loading at landing. The wing loading at takeoff was conditional upon the requirement that the aircraft be able to operate from a runway not longer than 9500 ft, under FAR conditions and noise limitations. Both aircraft meet this requirement, but it is interesting to note that the LH_2 design has the capability of operating from runways as short as 7800 ft.

The aircraft thrust-to-weight ratios (T/W), listed in Table 9 for sea level static conditions, were parametrically established as preferred values for each of the aircraft but for different reasons. For the LH_2 aircraft, the value of T/W which produced the minimum gross weight vehicle also provided the capability of meeting all of the design constraints such as takeoff distance, takeoff climb gradient, and noise limitations. The engines of the Jet A vehicle had to be sized to meet the Federal Air Regulation Part

TABLE 9

Comparison of Mach 2.7 Jet A and LH_2 Supersonic Cruise Vehicles (Mach 2.7, 234 PAX, Range = 4200 nmi)

	Units	Jet A	LH_2	Ratio Jet A/LH_2
Takeoff gross weight	lb	762,170	394,910	1.93
Operating empty weight	lb	317,420	245,235	1.29
Fuel weight				
Block	lb	330,590	85,395	3.88
Total	lb	395,750	100,675	3.93
Wing area	ft²	11,094	7,952	1.39
Wing loading				
Takeoff	lb/ft²	68.7	49.7	
Landing	lb/ft²	38.9	38.9	
Span	ft	133.5	113.0	1.18
Overall length	ft	297.0	340.2	0.87
L/D (cruise)		8.65	7.42	1.17
Specific fuel consumption (SFC) (cruise)	lb_{hr}/lb	1.501	0.575	2.61
Thrust per engine	lb	86,890	52,820	1.64
T/W (SLS)	—	0.456	0.535	
Federal air regulation takeoff distance	ft	9,490	6,080	1.56
Federal air regulation landing distance	ft	7,980	7,800	1.02
Landing approach speed (equivalent air speed)	kn	158.0	158.0	
Weight fractions	%			
Fuel		52.0	25.5	
Payload		6.4	12.4	
Structure		25.9	36.5	
Propulsion		9.9	15.1	
Equipment and operating items		5.8	10.5	
Energy utilization	Btu/seat nmi	6,189	4,483	1.38

36 flyover noise limit as the limiting condition. Although the LH_2-fueled aircraft has a higher T/W, its engines are still significantly smaller than those of the Jet A counterpart, because of its lower gross weight.

Energy utilization is a parameter used to compare the relative efficiencies of the subject aircraft in terms of the specific energy consumed in transporting the design payload the specified range. As noted, the subject supersonic transport aircraft carry 234 passengers plus full cargo (a total payload of 49,000 lb) for a distance of 4200 nmi at a cruise speed of Mach 2.7. With the LH_2-fueled aircraft, this required 4483 Btu per seat per nautical mile. The Jet A design requires 38% more energy, or 6189 Btu per seat per nautical mile. As stated in the basic guidelines in Table 6, both fuels were assumed to be available at the airport, so the energy expended in manufacturing, refining, or transporting them to the airport is not involved in this comparison.

Comparing the two aircraft on the basis of cost considerations, Table 10 presents development and production costs based on a production quantity of 600 aircraft and using 1973 dollars. The figures show a significant advantage for the hydrogen-fueled design. The LH_2 supersonic cruise vehicle (SCV) aircraft is almost $16 million cheaper than the comparable Jet A airplane in production, and development is estimated to cost $567 million less. Estimation of the price of an aircraft is a function of a number of parameters, the most dominant of which are its gross weight and the thrust of its

TABLE 10

Development and Production Costs — Mach 2.7
Jet A and LH₂ SCVs

	Costs in 10^6	
	Jet A	LH₂
Development		
Engine	1001	876
Airframe	3344	2902
Total	4345	3778
Production aircraft, each	61.4	45.5

engines. Both of these parameters are considerably lower for the LH_2 design, so that even using the higher cost factors which account for the anticipated increased complexity of manufacturing that design, there was a net saving.

Direct operating cost (DOC), as calculated according to the 1967 ATA equations,[12] is a function of the cost for the flight crew, fuel and oil, insurance, depreciation, and maintenance. Of these items, the LH_2 design has a decided advantage in insurance and depreciation, because of its lower production cost, and in maintenance, because of the factors noted earlier: lower gross weight and improved engine combustion conditions. The cost of the flight crew might be slightly lower for the Jet A aircraft because of the probable need for one member of the cabin crew to assume a higher level of responsibility in the isolated passenger compartment of the LH_2 airplane. Fuel cost is the largest, most important parameter in the list, and it can be expected to vary the most.

The effect of variations in cost of fuel on DOC for the subject Mach 2.7 supersonic cruise vehicles is shown in Figure 18. DOC, expressed in cents per available seat per statute mile, is plotted as a function of cost of both Jet A and LH_2 over a wide range of possible fuel prices shown along the abscissa in dollars per million Btu. For convenience, the price of Jet A is also presented across the top of the graph in the more familiar units of cents per gallon.

To provide a point of reference, in September 1975, United States international air carriers reported paying an average price of 36.6c/gal for Jet A fuel. The length of the horizontal dash line which joins the sloping lines representing the Jet A and the LH_2 vehicles, respectively, indicates airlines using LH_2 fueled aircraft could pay $1.55/$10^6$ Btu more for LH_2 without increasing their DOC. Due to the divergence of the sloping lines, as the cost of Jet A continues to increase, an even greater differential can be allowed for the cost of LH_2. For example, when Jet A costs 50c/gal, the "permissible" cost of LH_2 for parity in DOC is nearly $6/10^6$ Btu.

The basis of these DOC calculations is passenger load factor of 55%, aircraft utilization of 3600 hr per year, and aircraft service life of 50,000 hr. The following expression can be used to calculate the differential which can be paid for LH_2 over a specified price for Jet A, to provide parity in DOC for the subject Mach 2.7 SCVs.

$$\Delta C_{LH_2} = 0.335\, C_{JA} + 0.560$$

where ΔC_{LH_2} = cost increment in dollars per 10^6 Btu permitted for LH_2 to produce equal DOC and C_{JA} = cost specified for Jet A fuel in dollars per 10^6 Btu.

2.4.1.5. Environmental Factors

A comparison of some significant environmental acceptance parameters for the sub-

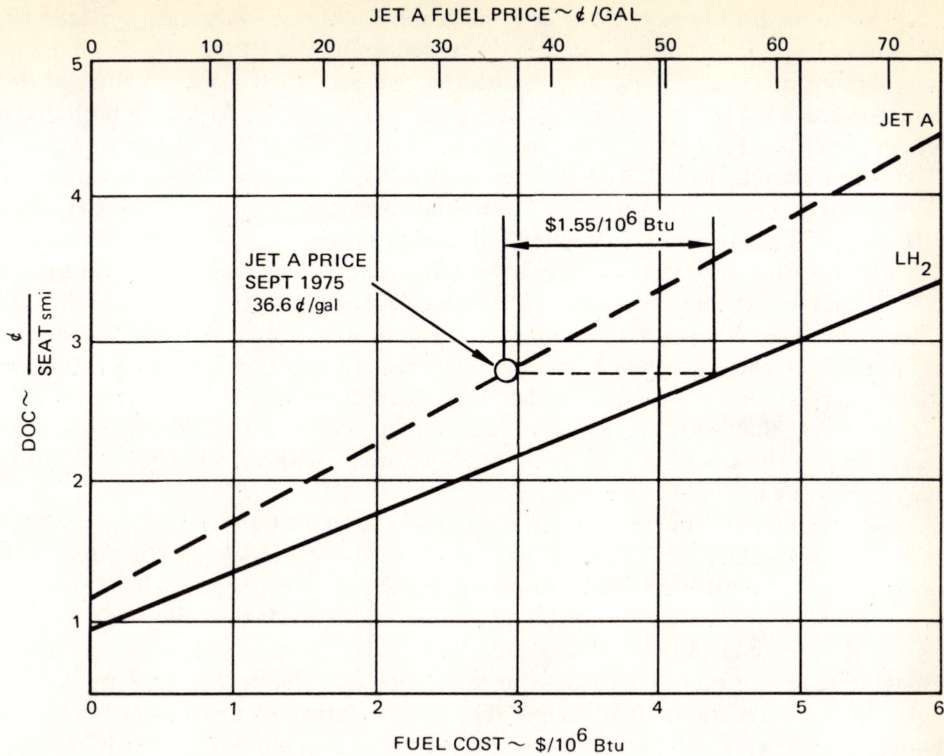

FIGURE 18.   Direct operating cost vs. fuel cost (Mach 2.7 SCVs).

TABLE 11

Environmental Parameters for Mach 2.7 SCVs

	LH₂	Jet A
Noise (EPNdB)		
Sideline, calculated (FAR 36)	104 (107)	108 (108)
Flyover, calculated (FAR 36)	102 (105)	108 (108)
Sonic boom overpressure (lb/ft²)		
Start of cruise	1.32	1.87
End of cruise	1.19	1.40
Maximum encountered	2.08	2.50
(during climbout)		
Exhaust emissions		
NO_x (g/kg)	Low	3.7[a]
CO (g/kg)	None	90.0[a]
UHC (g/kg)	None	0.5[a]
Odors	None	Objectionable
H_2O (lb/nmi)	153	76.9

[a] Data for GE-J85 engine, simulated flight at Mach 1.6, 55,000 ft.

ject supersonic transport aircraft is shown in Table 11. Noise generated by the turbofan engines during takeoff is listed for the two standard measurements, sideline and flyover. Sideline noise was calculated to determine the maximum Effective Perceived Noise Level (EPNL) in units of decibels (EPNdB) at a locus of points 0.35 nmi from the centerline of the runway, extending the length of the runway and projecting beyond. Flyover noise is a similar calculation to determine the EPNL at a point 3.5 nmi from the point of brake release in line with the center of the runway. Values of these parameters are listed for the subject aircraft using the two fuels. Although both designs meet the prescribed FAR Part 36 specification, the LH_2 aircraft is considerably quieter. In fact, the LH_2 vehicle is 2.8 dB quieter than the FAR 36 specification requires for an aircraft of its design gross weight. Compared to the Jet A airplane, the LH_2 design is 4 dB quieter in sideline and 6 dB quieter in flyover noise.

The LH_2 airplane also offers appreciable reduction in overpressures caused by sonic boom. The table lists sonic boom overpressure calculated for three discrete points during the flight. The values for the hydrogen airplane are lower primarily because that aircraft is lighter and has a much smaller wing area. In addition, during climbout and at start of cruise, it is somewhat higher than the equivalent Jet A design. It is not likely, however, that with the overpressures shown, even an LH_2-fueled SCV would be permitted to fly supersonically over inhabited areas without significant additional reduction in sonic boom overpressure.

Concern about pollution of the environment was one of the principal causes for cancellation of the U.S. SST development program in March 1971. The concern was for pollution of the stratosphere from products exhausted during supersonic cruise as well as pollution in the vicinity of the airport resulting from operations on and near the ground. Engine exhaust products deposited in the stratosphere are a problem because their residence time is measured in terms of years due to the very low levels of circulation characteristic of that region. The accumulation of noxious products from operation of a fleet of SSTs over a number of years can, therefore, reach enormous totals, even though the amount emitted per vehicle may be relatively small. The effect of SST exhaust products in the vicinity of airports is of interest because it is additive to the pollution which already exists around centers of population. The addition simply makes a bad situation worse.

Typical exhaust products from combustion of Jet A fuel and air in aircraft gas turbine engines include carbon; nitrogen; various hydrocarbons; oxides of carbon, nitrogen, and sulfur; and H_2O. Although modern designs of aircraft gas turbines have reduced soot and unburned hydrocarbons to reasonable levels under cruise conditions, NO_x remains a problem. During periods of off-design operation of the engines, viz., during taxi and hold on the ground and during approach, landing, and takeoff, all of the noxious products (carbon monoxide, hydrocarbons, sulfides and sulfates, and oxides of nitrogen) are of concern.

On the other hand, the exhaust products which result from combustion of hydrogen and air are limited to water vapor (H_2O), nitrogen (N_2), and some oxides of nitrogen (NO and NO_2). The oxides of nitrogen (NO_x) are the only products which would be of concern around an airport and, as is developed later, there is a very real expectation that production of those materials can be substantially reduced, relative to the quantities produced by corresponding Jet A fueled engines.

Table 11 lists exhaust products which are representative of those from Jet A fueled turbojet engines during cruise flight for comparison with principle emissions which might be expected from LH_2 fueled engines. Values of the first three items — oxides of nitrogen (NO_x), carbon monoxide (CO), and unburned hydrocarbons (UHC) — are shown as reported from tests on a Jet A fueled GE-J85 turbojet during simulated flight at Mach 1.6 at an altitude of 55,000 ft. Obviously, if LH_2 is used there will be no CO or UHC.

The principal exhaust product from LH_2-fueled aircraft will be water vapor. Water vapor (H_2O) emission data in Table 11 were calculated for both fuels as the total which would be produced per nautical mile during cruise flight by all four engines on the respective aircraft of Table 9. During cruise, the LH_2 airplane uses 17.1 lb of hydrogen fuel per nautical mile. Assuming 100% conversion of H_2 to H_2O in the exhaust stream, this generates 153 lb of H_2O per mile. By contrast, the equivalent design of Jet A fueled SCV will use 61.4 lb of hydrocarbon fuel per nautical mile and, again assuming 100% conversion of the hydrogen in the exhaust, this will generate 76.9 lb of H_2O per mile.

The emission of objectionable odors from the Jet A fueled engines is noted for particular reference to operation on and near airports. There is no odor from LH_2-fueled engines.

No firm data exist to serve as a basis for accurately calculating the amount of NO_x which might be produced from carefully designed LH_2 engines. There is little experimental data available which reflect what might be achieved with a combustor specifically designed to burn hydrogen. However, theory indicates that such emissions can be minimized by proper design of the H_2/air injector and combustor. In addition, Table 9 shows that the engine thrust requirement is reduced by almost one half for the LH_2 fueled aircraft. This thrust reduction further decreases the total NO_x emissions per flight.

Use of hydrogen for fuel in the SCV would greatly alleviate pollution of the environment near the ground during off-design operation of the engines. The following is a brief discussion of the potential of the hazards to the environment posed by exhausting the noxious products from each fuel system into the stratosphere during cruise.

Ozone, normally present in the stratosphere, serves beneficially to shield the earth from ultraviolet radiation. Analyses have shown that oxides of nitrogen can act as catalysts to reduce ozone (O_3) to oxygen (O_2) in an irreversible reaction, thereby reducing the availability of ozone to perform its useful function.[14] Because it is a catalytic reaction, a small amount of NO or NO_2 can drastically influence the ozone concentration in the stratosphere before it is finally "washed" into the troposphere by normal circulation.

Oxides of nitrogen are formed during the combustion process by reaction of nitrogen and oxygen from the air at high temperatures. The reaction increases exponentially with temperature and linearly with dwell time. It is, therefore, paramount in the interest of minimizing NO_x that combustion temperature be kept as low as is consistent with acceptable engine performance and that the duration of the combustion process be as short as possible. Ideally, the maximum temperature reached in the combustion chamber would be the turbine inlet temperature, e.g., about 3000°R for advanced design engines. With conventional hydrocarbon fuel, the temperature profile across the combustor is very irregular; local peak temperatures as high as about 4600°R are reached during combustion of a droplet of liquid fuel, due to nonuniform distribution and mixing of the fuel and air as the droplet decreases in size. It follows, therefore, that if combustion could be caused to occur uniformly throughout the combustion zone (i.e., no local hot spots at temperatures above 3400°R), the problem of excessive NO_x formation would be greatly reduced.

This ideal situation can be approached if the fuel is injected in gaseous form instead of as a spray of liquid droplets. Mixing of gaseous fuel with air can be so uniform and complete that local pockets of excessively lean or rich mixture can be largely eliminated. Experience with burning both methane (CH_4) and hydrogen in gaseous form in gas turbine engines has demonstrated the smooth, uniform, and complete combustion possible with this situation. In addition, the very rapid mixing, diffusion, and combustion characteristics of hydrogen have been shown by Pratt & Whitney® Aircraft to permit considerable shortening of the combustion chamber, thus reducing the dwell time of the reaction and also benefiting the size and weight of the engine, without compromising the efficiency of process.[6]

Water vapor in the stratosphere can affect the atmospheric heat balance in several ways.[9] Clouds of water vapor or ice crystals can affect the reflection and absorption of both ultraviolet radiation from the sun and infrared radiation emitted from the earth. In addition, water vapor in the stratosphere can react with free oxygen atoms (O), thus reducing their availability to react with molecular oxygen (O_2) in ultimately forming ozone (O_3).

As previously stated, a hydrogen-fueled SCV will generate approximately twice as much water vapor per mile as an equivalent vehicle fueled with Jet A. However, it is the consensus of authorities that water vapor added to the stratosphere by operation of a fleet of SSTs would have a negligible effect on atmospheric temperatures, simply because the amount so added would constitute a small percentage of that continuously present in the stratosphere as a result of natural occurrences (thunderstorms, Hadley cell circulation, and natural oxidation of methane).[15] In a calculation to investigate an extreme example, Manabe and Wetherald have estimated a possible increase in the temperature of the surface of the Earth of no more than 1°F on the basis of the very conservative assumption that water vapor flux from aircraft flying in the stratosphere is the same order of magnitude as the natural flux transferred from the troposphere.[16]

Jet A fuel can have only 0.3% of sulfur without exceeding the specification limit. However, in 1971, those actively concerned about the effect a large number of SSTs would have on climatic conditions on earth reasoned that the sulfur in the fuel could react with oxygen from the air during combustion to form SO_2 and SO_3. The SO_3 in the exhaust would then attract a free oxygen atom to form the sulfate radical, SO_4. This would occur in the form of an aerosol: a suspension of fine liquid droplets. The final step in the reasoning was that, if such particulates reached high concentrations in the stratosphere, they could cause severe earth surface temperature changes similar to those engendered by cataclysmic volcanic eruptions.

Obviously, this problem would not exist if the fuel were free of sulfur as a contaminant, as would be the case if LH_2 was used as the fuel. A large part of the sulfur can be removed from petroleum fuels by hydrogenation so that the production of sulfates in the atmosphere can be minimized. It is feasible to remove up to about 80% of the sulfur content of Jet A fuel in this manner, at extra cost of course. This would minimize the conjecture about the problem. If hydrogen is used, the problem would be eliminated, even as a subject for speculation.

The question that was raised concerning the effect the CO_2 exhausted from SST engines might have on the environment of Earth was similar to that discussed above for sulfur compounds. In this case, the gaseous CO_2 was postulated to cause a "greenhouse" effect wherein ultraviolet solar radiation to the earth would be accepted but reradiation of infrared energy would be restricted by the blanket of CO_2 in the stratosphere. Most responsible authorities have agreed that the maximum predictable long-term increases in CO_2 due to SST fleet operations would almost certainly cause no significant change in the temperature of the surface of the Earth. The principal problem with the subject is that it continues to be raised as a demagogic issue. The use of hydrogen fuel would eliminate the argument completely.

A summary of the situation as currently seen for pollution of both the stratosphere and the atmosphere at ground levels for SSTs using both hydrocarbon fuel and hydrogen fuel is presented in Table 12. It is concluded that use of hydrogen fuel would tremendously reduce the problems which were responsible, in large measure, for stopping the previous U.S. SST development program.

TABLE 12

Summary of Pollution Problems for Supersonic Transports

Pollutant	Area of concern	Fuel systems	
		Hydrocarbon fuel	Hydrogen fuel
Near the ground			
Exhaust products emitted on and near airports	Pollution of atmosphere near cities	Unburned hydrocarbons, carbon monoxide, sulfur oxides and oxides of nitrogen are produced; no significant improvement in total noxious emissions expected	The only noxious product produced, NO_x, can probably be reduced to acceptable levels
In stratosphere			
NO_x	Represents a serious potential hazard to ozone content in stratosphere thereby affecting radiation spectrum to earth; requires a solution which will prevent formation of NO_x in the combustion process	A design solution is theoretically possible; requires extensive R&D effort; solution will probably involve penalties in weight and cost	NO_x can probably be reduced to acceptable levels without significant penalty
H_2O	A relatively minor threat to ozone content in stratosphere	Acceptable as is	Acceptable as is
H_2O	Formation of clouds of ice crystals in stratosphere; of uncertain concern	Probably acceptable as is; needs verification	Probably acceptable as is; needs verification
Sulfur compounds	Concentration of particulate matter in stratosphere may cause temperature changes on earth surface	Additional refining of fuel required to reduce sulfur content; adds to cost of fuel	No sulfur
CO_2	Potential "greenhouse effect"; of negligible concern	No solution possible, but problem not significant	No CO_2

2.4.2. Subsonic Transport Aircraft

The potential of using hydrogen as the fuel for subsonic commercial transport aircraft has been investigated for several passenger and cargo payload-range combinations.[17,18] Only the passenger transport application will be discussed here. Although there were differences in design of the fuel containment systems and in their location within the aircraft, there were no significant differences in the findings resulting from the comparison of efficient LH_2-fueled aircraft designs with comparable Jet A fueled aircraft which were functions of the type of payload carried. (For those interested in the specifics of the design or performance characteristics of the cargo aircraft, see Reference 17.)

As was done for the supersonic aircraft, conventionally fueled Jet A aircraft designs were established in accordance with common guidelines to provide a valid basis for reference in studying the advantages or disadvantages of using LH_2 in subsonic transport aircraft.

Figure 19 is a picture of a model of a preferred configuration of LH_2-fueled, 400-passenger aircraft designed to fly 5500 nmi at a cruise speed of Mach 0.85. The general arrangement of the design is illustrated in Figure 20. Externally, there is little to distinguish the configuration from current, conventionally fueled, wide-body transports. Internally, the general layout of the passenger compartment, relative to the fuel tanks, is similar to that described previously for the supersonic transport aircraft, i.e., the passengers are located in the central portion of the fuselage is a double-deck arrangement with the fuel tanks located forward and aft. There is no provision for physical access between the passenger compartment and the flight station.

In the subsonic design, the fuselage is circular in cross-section with a lower lobe provided for cargo. The passengers are seated in a 2-4-2 arrangement on both decks for a total of 16 seats per row. Stairs are provided at both ends of the cabin for interdeck access. Cargo, amounting to 10% of the design passenger load, is carried in space provided below the passenger compartment. Total payload weight for the 400-passenger design is 88,000 lb.

The wing incorporates high-lift devices including 15% leading-edge slats and 35%

FIGURE 19. LH_2 subsonic passenger transport.

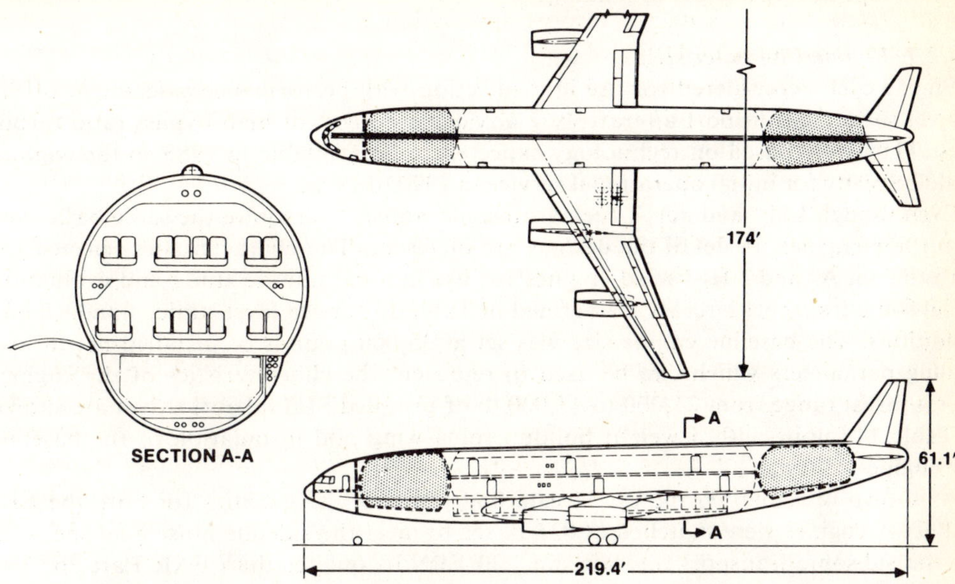

FIGURE 20.   General arrangement, LH₂ subsonic passenger transport.

double-slotted Fowler flaps out to the outboard engine. Conventional ailerons are fitted to the outboard wing panel. Spoilers are used in flight for direct lift control and for deceleration during landing ground run.

Except for differences in detail, the conceptual design of the hydrogen tanks and their thermal protection system for the subsonic aircraft are the same as described for the LH₂-fueled supersonic transport aircraft. For example, the tanks are the integral type. However, in the subsonic aircraft, 6 in. of rigid, closed-cell, plastic foam is applied to the outside of the aluminum tank to limit hydrogen boil-off to an acceptable rate. The extra thickness in the subsonic case was required because of the longer flight time of the design mission. The foam, in turn, is covered with a thin plastic which serves as a secondary vapor barrier to prevent cryopumping of air in the event the rigid plastic foam insulation develops minute cracks during aircraft service. Build-up of frost on the outer surface of the insulation when the airplane is on the ground will not be a problem because the minimum temperature reached with 6 in. of foam will be about 46°F on a standard day.

Since a high-temperature heat shield is not required for the subsonic aircraft, the outer covering can consist of a simple fiberglass or composite cover wrapped around the entire tank assembly to provide mechanical protection for the vapor barrier and the plastic foam insulation. This outer cover also serves as the skin of the aircraft in the tank areas and provides aerodynamic fairing over the truss framework joining the tanks to the conventional fuselage structure. In the event either the foam insulation or the tank structure requires repair, the fiberglass cover could simply be cut away locally and then patched upon completion of the work. Inspection of the tank structure can be accomplished from inside the tank. A crawl hole would be provided for the purpose.

Further mechanical protection is provided on the bottom of both fuel tanks, forward and aft of the cargo compartment. This protection is 18 to 24 in. of energy-absorbing aluminum honeycomb supported from the tank bottom. It provides protection to the fuel tanks from foreign object damage and from the effects of possible over-rotation or tail scrape during takeoff or landing.

2.4.2.1. Engine Characteristics

Engine cycles considered for use in connection with performance assessment of the subject subsonic transport aircraft were advanced designs of high bypass ratio turbofans. They were based on technology expected to be available in 1985 so the engines could be ready for initial operational service in 1990 to 1995.

Even though LH_2- and Jet A-fueled subsonic transports require the same basic type of turbine engines, in detail the designs are different. The characteristics selected for both the Jet A- and LH_2-fueled engines for use in a parametric study and evaluation of subsonic transport aircraft are defined in Table 13 for sea level static, standard day conditions. The baseline engine size was set at 35,000 pounds of installed net thrust. Scaling parameters which can be used to represent the characteristics of the engines over a thrust range from 25,000 to 45,000 lb of installed SLS takeoff thrust are shown in Table 14, along with a weight buildup for a wing pod installation of the baseline size engine.

As shown in Table 13, the fan stream and primary jet velocities for both the LH_2 and Jet A engines were matched at 835 ft/sec to meet the sideline noise goal specified for the subsonic transport aircraft, i.e., 20 EPNdB quieter than FAR Part 36. The bypass ratio (BPR), at SLS, is 12.95 for the LH_2-fueled engine and 10.9 for the Jet A. For the cruise condition the BPR values are 13.5 and 11.6.

The difference in bypass ratios stems directly from the difference in methods used to cool the high-pressure (HP) turbines of the two engines. The method selected for the LH_2 engine was to use an intermediate coolant in a closed loop with a hydrogen-to-coolant heat exchanger. The Jet A-fueled engine is constrained to use bleed air from its compressor. Since ambient air compressed to a pressure ratio of 35:1 will be heated to a temperature of nearly 1000°F, there is little heat sink available to do much effective cooling. Accordingly, 5% of the compressor air flow was considered a reasonable amount to accomplish necessary cooling of HP turbine vanes and blades for the selected turbine inlet temperature of 2850°F.

Because the air used to cool the turbine parts is exhausted from the vanes and blades after passing through the item being cooled, there is a loss in hydrodynamic efficiency due to disturbance of the flow of the turbine working fluid. Since the LH_2 engine does not use air for cooling, its turbine efficiency was rated 1% higher than that of its conventionally fueled counterpart: 91 vs. 90%. In addition, more cycle energy is required of the Jet A engine gas generator to account for the work performed in compressing the air used for cooling.

A mixture of sodium (Na) and potassium (K) called Nak, such as has been used successfully for cooling in some nuclear reactor designs, was assumed for the LH_2 engine turbine. This is the method assumed as a basis for the engine performance calculated for the supersonic aircraft. It should be noted, however, that other methods of cooling the high-pressure turbine are available, some of which might be less technologically advanced and therefore less risky, although they would also be expected to be less advantageous in performance. An example is where air bled from the compressor is cooled in a hydrogen air heat exchanger before it is routed to the turbine. As a result of using cold air to do the vane and blade cooling instead of hot air direct from the compressor, less air is required so there is less disturbance of the turbine working fluid and the turbine efficiency will be higher. This approach is currently being evaluated in a study by Lockheed for NASA-Langely Research Center.[19]

TABLE 13

Baseline Engine Characteristics for Subsonic Aircraft (Sea Level Static, Standard Day Conditions)

Overall characteristics	Hydrogen fueled	Hydrocarbon fueled
Installed net thrust	35,000 lb	35,000 lb
Installed SFC	0.094 (lb/hr)/lb	0.281 (lb/hr)/lb
Turbine inlet temperature	3040°R	3040°R
Bypass ratio	12.95	10.90
Overall pressure ratio	35.0	35.0
Jet exhaust velocity[a]	835 ft/sec	835 ft/sec
Fan design		
Stages	1	1
Airflow — $Wa\sqrt{\theta}/\delta$	1360 lb/sec	1356 lb/sec
Pressure ratio	1.51	1.51
Polytropic efficiency	91%	91%
Diameter	84.4 in	84.4 in.
Tip velocity	1390 ft/sec	1390 ft/sec
Fan face Mach number	0.56	0.56
Hub/tip ratio	0.35	0.35
Compressor design		
Compressor pressure ratio	23.3	23.3
Polytropic efficiency	92.0%	92.0%
Airflow	96 lb/sec	113 lb/sec
Combustor		
Efficiency	100%	100%
Total pressure loss	4.5%	4.5%
High-pressure turbine		
Pressure ratio	3.9	4.6
Stages	2	2
Adiabatic efficiency	91%	90%
Cooling air	0	5%
Low-pressure turbine		
Pressure ratio	7.4	6.2
Stages	4	4
Adiabatic efficiency	91%	91%
Cooling air	0	0
Nozzle design		
Configuration	Coplaner, fixed convergent nozzle	Same
Core C_v	0.995	0.995
Fan C_v	0.995	0.995
Acoustic treatment		
Inlet	Variable geometry throat — throat Mach > 0.8 during takeoff and approach, inlet wall treatment	Same

[a] Core and fan exhaust velocities are equal at SLS.

TABLE 13 (continued)

**Baseline Engine Characteristics for Subsonic Aircraft (Sea Level
Static, Standard Day Conditions)**

Overall characteristics	Hydrogen fueled	Hydrocarbon fueled
Fan duct exhaust	All treatment on both core engine and outer wall, one treated duct ring	Same
Core exhaust	Wall treatment	Same
Nacelle geometry		
Maximum diameter	106 in.	Same
Overall length	283 in.	Same
Inlet highlight diameter	86 in.	Same
Inlet throat diameter	78 in.	Same
Cruise throat Mach number	0.73	Same

The cryogenic hydrogen of the LH_2-fueled aircraft provides a large heat sink which can be used effectively in one way or the other for turbine cooling. Unconstrained by other limits, the LH_2-fueled engine's compression ratio and turbine inlet temperature can be independently selected for the optimum overall propulsion performance. However, practical limits of overall compression ratios and turbine inlet temperatures appear to be 40:1 and 3800°R, respectively, for engines which might be started in development in 1985. Ultimately, selection of the best cycle parameters for subsonic transport aircraft should be based upon a parametric evaluation related to mission requirements. For purposes of the initial evaluation of LH_2-fueled subsonic aircraft, reported in Reference 17, conservative values of 35:1 and 3040°R, as shown in Table 13, were selected for overall compression ratio and turbine inlet temperature, respectively.

The engines are controlled by scheduling the compressor rotor speed as a function of turbine inlet temperature (TIT) to provide a relatively constant corrected gas flow to the low-pressure turbine. The fan speed is allowed to balance out in the resulting cycle condition. The only limiter used is a maximum fan speed of 108% which affects a few high-altitude, low-Mach number flight conditions.

Because the aircraft takeoff design condition was specified to be from an airport at an elevation of 1000 ft on a hot day (90°F), the engines were flat-rated to provide full thrust at this condition.

2.4.2.2. Aircraft Characteristics

A parametric design study was conducted to determine preferred configurations of subsonic transport aircraft using both LH_2 and Jet A fuels. By proper choice and variation of input parameters, aircraft designs were evolved which not only performed the specified missions and met all design constraints, but also provided minimum direct operating cost. The Jet A airplane design has a fuselage of about the same fineness ratio (length-to-diameter ratio) as its LH_2 counterpart. Since the Jet A version carries all of its fuel in the wing box in conventional fashion, the passenger compartment can occupy the entire usable length of the fuselage. Accordingly, it is not necessary to provide as many seats per row and the fuselage can be a smaller diameter. The Jet A

TABLE 14

Installed System Weight and Scaling Data for Baseline Turbofan

Item	lb
Bare engine	5380
Accessories and gearbox	480
Inlet, variable geometry	1005
Mounting brackets and pylon splitter fairing	200
Nacelle	990
Gas generator cowl and tail pipe	510
Fan duct acoustic ring	275
Thrust reverser	620
Total pod weight per engine	9460

Nacelle Scaling Data

$$\text{Wt POD} = \text{WT}_{\text{POD(REF)}} \left(\frac{F_{N_{SLS}}}{F_{N_{SLS\,(REF)}}} \right)^{1.07}$$

$$\text{Dia} = \text{DIA}_{\text{(REF)}} \left(\frac{F_{N_{SLS}}}{F_{N_{SLS\,(REF)}}} \right)^{0.50}$$

$$\text{Length} = \text{LENGTH}_{\text{(REF)}} \left(\frac{F_{N_{SLS}}}{F_{N_{SLS\,(REF)}}} \right)^{1.07}$$

Note: Baseline thrust = 35,000 lb (SLS, installed).

TIT = 3040°F, OPR = 35.0, FPR = 1.51.

aircraft has a cabin diameter of 230 in. and uses a 2-2-2 seating arrangement on each deck for a total of 12 seats per row. Corresponding numbers for the LH_2 design are 261 in., a 2-4-2 seating arrangement per deck, for 16 seats per row.

A comparison of several key design and performance parameters is presented in Table 15 for evaluation of the relative merits of the Jet A- vs. the LH_2-fueled subsonic transport aircraft. The comparison is much the same as previously observed for the supersonic transport designs in Table 9; however, in the present case, the differences are smaller because there is less fuel involved. The fundamental advantages gained from using hydrogen fuel in aircraft stem from substituting a high-energy fuel for a relatively lower-energy fuel. In those airplane designs where the mission does not require much fuel, it is apparent that there will be less performance advantage to switching fuels.

Although the advantages of using LH_2 fuel in place of Jet A in the subject subsonic

TABLE 15

Comparison of Jet A-vs. LH$_2$-Fueled Subsonic Passenger Aircraft (Mach 0.85, 400 PAX, Range = 5500 nmi)

	Units	Jet A	LH$_2$	Ratio (Jet A/LH$_2$)
Takeoff gross weight	lb	523,200	391,700	1.34
Operating empty weight	lb	244,400	242,100	1.01
Fuel weight				
Block	lb	165,500	52,900	3.13
Total	lb	190,800	61,600	3.10
Wing area	ft^2	4,186	3,363	1.24
Wing loading				
Takeoff	lb/ft^2	125	116.5	1.07
Landing	lb/ft^2	106	101	1.05
Span	ft	194.1	174	1.12
Fuselage length	ft	197	219	0.90
L/D (cruise)	—	17.91	16.07	1.12
Specific fuel consumption (cruise)	$^{lb}/_{hr}$/lb	0.581	0.199	2.92
Thrust per engine (sea level static)	lb	32,700	28,700	1.14
T/W (sea level static)	—	0.25	0.293	0.85
Federal air regulation takeoff distance	ft	7,990	6,240	1.28
Federal air regulation landing distance	ft	5,210	5,810	0.90
Approach speed (equivalent air speed)	kn	124	135	0.92
Weight fractions	%			
Fuel		36.5	15.7	2.23
Payload		16.8	22.5	0.75
Structure		26.0	30.7	0.85
Propulsion		6.4	12.3	0.52
Equipment and operating items		14.3	18.8	0.76
Energy utilization	$\dfrac{Btu}{seat\ nmi}$	1,384	1,239	1.12

transport aircraft are not as dramatic as those observed in Table 9, they are nevertheless quite significant. In Table 15, the conventionally fueled Jet A design is seen to be 34% heavier in gross weight, to require over 3 times the weight of fuel, and to need engines that deliver 14% more thrust than its counterpart designed to use hydrogen fuel. There is negligible difference in operating empty weight of the two aircraft, but the LH$_2$ design has the capability of operating from shorter runways, and it would be easier to handle on the ground in the terminal area because of its shorter wing span. The weight fractions listed in the table are referenced to the takeoff weight of the aircraft.

A factor of particular interest is the comparison of energy utilization. This is the amount of energy expended in performing the design mission, expressed in terms of Btu per seat per nautical mile. The Jet A airplane requires 12% more energy to transport 400 passengers, 5500 nmi, at Mach 0.85 cruise speed than does the corresponding LH$_2$-fueled aircraft.

The critical design condition for the Jet A subsonic aircraft was the requirement to operate from an 8000-ft runway. Takeoff within this distance established a limit on the combination of thrust to weight ratio (T/W) and wing loading (W/S) which could be used in the parametric analysis to determine the preferred aircraft design, i.e., that which provided minimum direct operating cost (DOC). The LH_2-fueled airplane design selection was influenced only by the requirement not to exceed an equivalent approach speed of 135 kn. This determined the wing loading for the LH_2 airplane. Because of its lower gross weight, the wing area of the LH_2 airplane could be smaller. This resulted in a correspondingly less effective span for high-lift devices, which in turn meant that the aircraft needed a lower wing loading at landing than did the Jet A-fueled design.

Engine size selection for the LH_2 design was not constrained by any of the design guidelines so choice of T/W could be made on the basis of the value which provided minimum DOC. As a result of the higher T/W of the LH_2 design, that aircraft is able to take off from a runway 1750 ft shorter than its Jet A counterpart.

The fundamental reason for the improved performance of the hydrogen-fueled design is the favorable tradeoff of its lift-to-drag ratio (L/D) and its specific fuel consumption (SFC) in cruise, relative to the equivalent Jet A-fueled design. Although the LH_2 design has an L/D which is 10% lower due to the low density and cryogenic nature of hydrogen, the much more favorable operating value of its SFC more than compensates. For an explanation of the reasons for this, see the discussion presented for the supersonic transport aircraft relative to Table 9.

A graphic comparison of the relative sizes of the subject subsonic transport aircraft is shown in Figure 21. The significantly smaller wing area and shorter span of the LH_2-fueled design are readily apparent, as are the smaller diameter and shorter fuselage length of the Jet A airplane.

Table 16 is a summary of development and production costs calculated for the subject aircraft. The LH_2 design costs slightly more, both to develop and to produce, than

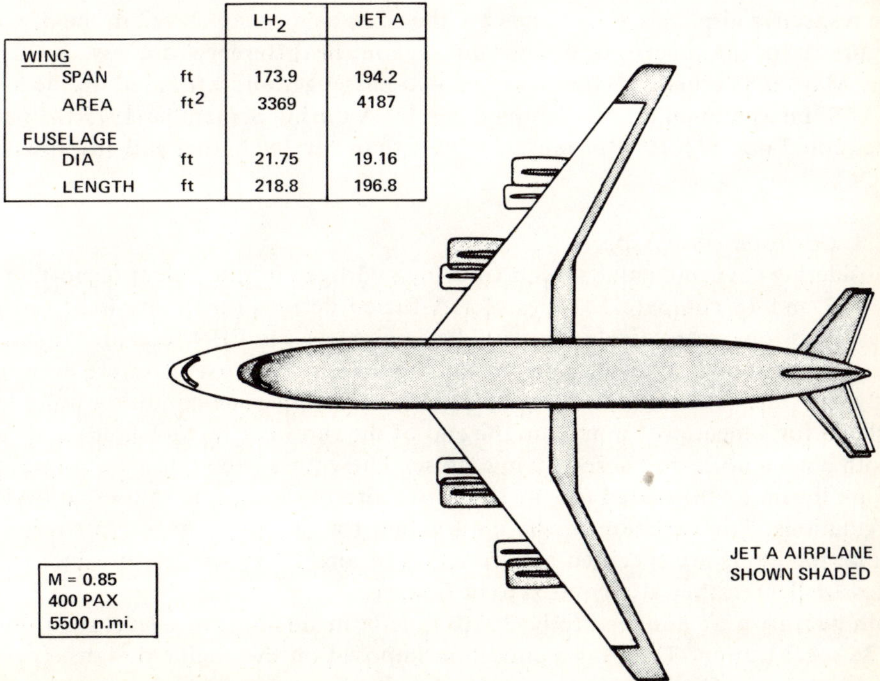

FIGURE 21.   Size comparison: LH_2 vs. Jet A subsonic passenger aircraft.

TABLE 16

Development and Production Costs — Jet A and
LH$_2$ Subsonic Passenger Aircraft (Mach 0.85,
400 PAX, Range = 5500 nmi)

	Costs in $10^6	
	Jet A	LH$_2$
Development		
Airframe	692.5	669.5
Engine	416.6	455.0
Total	1109.1	1124.0
Production		
Airframe	19.86	20.10
Engine	3.29	3.50
Avionics	0.50	0.50
R&D amortization[a]	2.81	2.82
Total	26.46	26.92

[a] Based on 350 aircraft and 2000 engines.

its Jet A counterpart. Production costs of both aircraft are based on amortizing development expenses over 350 airplanes and 2000 engines. It should be noted that the cost of basic hydrogen technology development was assumed to be funded separate and apart from the traditional aircraft development costs represented in the table. As discussed later, this is a 6-year program estimated to cost approximately $76 million which is suggested for implementation before a decision need be made to proceed with development of a commercial transport airplane. The cost of this basic technology development is not included in the costs shown in Table 16.

Figure 22 is an illustration of the effect the price of each of the fuels has on DOC of the respective airplanes. The trends are the same as those observed in the discussion of Figure 18 for the supersonic designs but, again, the differences are less. In this case, for the Mach 0.85 cruise, 5500-nmi range, 400-passenger aircraft based on the average price U.S. international air carriers paid for Jet A during September 1975 (36.6¢/gal), airlines could pay 52¢/10^6 Btu more for LH$_2$ than for Jet A fuel and still break even on DOC.

2.4.2.3. Environmental Aspects

Considering environmental aspects of using hydrogen in subsonic transport aircraft, Tables 17 and 18 compare LH$_2$- vs. Jet A-fueled designs on the basis of noise and exhaust emissions, respectively. Table 17 lists noise levels in EPNdB. Three operational conditions are shown: flyover, sideline, and approach. The first two were defined previously; they are evidence of engine noise only. The approach condition noise level is calculated for a location 1 nmi from the end of the runway (the threshold) and considers both engine noise and aerodynamic noise. The values shown in parentheses in the table are the limits calculated for the respective aircraft according to present FAR Part 36 regulations. The variation of the limit values for the different aircraft reflects the fact that the limits are specified as a function of aircraft gross weight. Since the LH$_2$ designs are lighter, they are required to be quieter.

Both aircraft were designed to be 20 dB quieter in sideline noise level than the FAR Part 36 specification. This was a condition imposed on the engine designs, regardless of the fuel used. The different noise level reductions calculated for the aircraft for

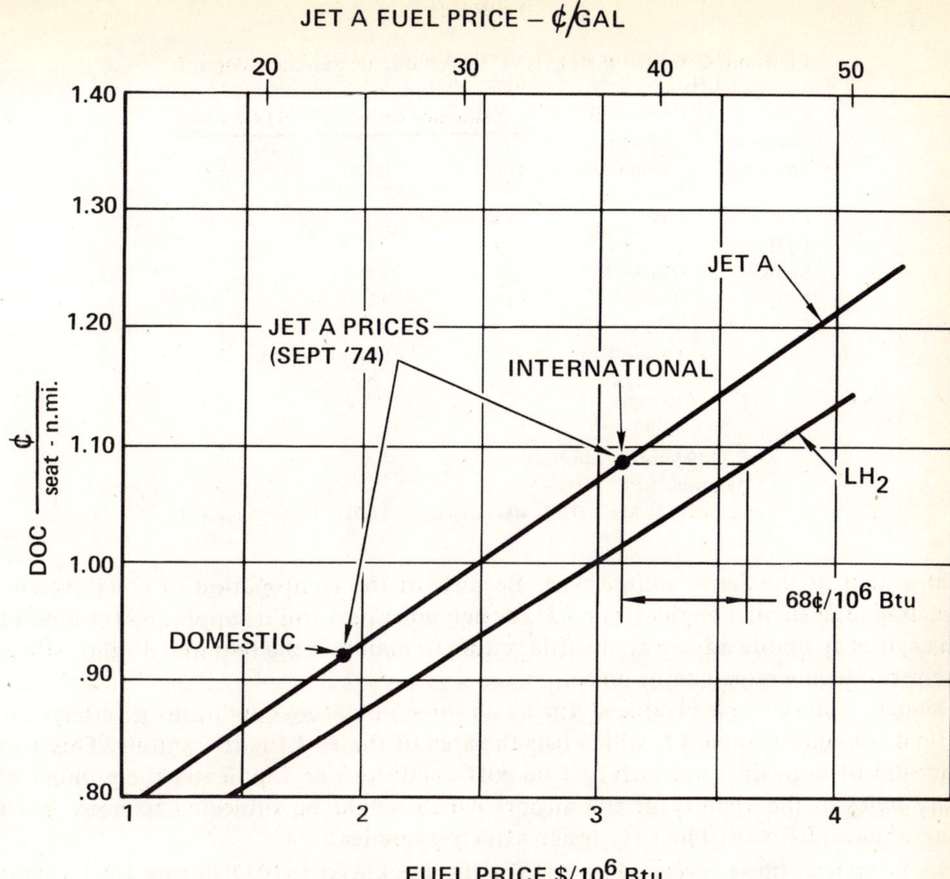

FIGURE 22.   Sensitivity of DOC to fuel cost — long-range subsonic aircraft.

TABLE 17

Noise Comparison of LH₂ vs. Jet A Subsonic Passenger Aircraft (Mach 0.85, 400 PAX, Range = 5500 nmi)

Aircraft	Noise levels (EPNdB)			Area of 90 EPNdB contour (mi²)
	Flyover	Sideline	Approach	
LH₂	89.2 (104.9)	87.2 (106.8)	98.4 (106.8)	4.3
Jet A	94.2 (107)	87.8 (107.6)	96.7 (107.6)	4.7
L-1011 (certification tests)	96.0 (105.6)	95.0 (107)	102.8 (107)	6.6

Note: ( ) = Federal Air Regulation Part 36 limits.

flyover and approach thus reflect the effect of different T/W and L/D of the aircraft. It may be noted in the table that there is one situation where the LH₂-fueled aircraft is slightly noisier than its Jet A-fueled counterpart, i.e., during approach. This results from a combination of characteristics. The LH₂ aircraft has small engines because of its lower gross weight. It has a lower L/D, and its weight at landing is approximately

TABLE 18

Emissions Comparison of LH_2 vs. Jet A Subsonic Passenger Aircraft

Emission product	Engine condition	Estimated emission level (g/kg fuel)		
		Goal	Jet A	LH_2
CO	Idle	14	30	0
UHC	Idle	2	4	0
Smoke	Takeoff	25[a]	15[a]	0
NO_x	Takeoff	13	12	$\leq 12^b$
H_2O	Cruise		41.9 lb/nmi[c]	82.4 lb/nmi[c]
Odors	Ground operations		Objectionable	None

[a]   SAE 1179 Smoke Number
[b]   g/ (kg fuel/2.8)
[c]   Calculated for Mach 0.85, 400 passenger, 5500 mm range aircraft

equal to that of the Jet A counterpart. Because of the combination of equal weight, more drag, and smaller engines, the LH_2 design must perform its approach for landing with engines at a more advanced throttle setting to maintain the required 3° glide slope, thereby producing more engine noise.

The net result of noise produced during all three operating conditions is reflected in the final column of Table 17, which lists the area of the 90-EPNdB contour. This area is the sum of both the approach and takeoff conditions and indicates the number of square miles in the vicinity of the airport which would be subjected to noise levels greater than 90 EPNdB. The LH_2 design affects a smaller area.

For reference, noise levels measured for the Lockheed L-1011 during FAA certification tests are also listed in Table 17. Since this transport is the quietest U.S. wide-body airliner in operation today, it is clear that significant improvement in transport aircraft of the future can be expected, particularly if LH_2 is used as the fuel.

Table 18 presents a comparison of the principle emissions which will be produced by engines from advanced designs of subsonic transport aircraft. Goals specified by NASA for the study of Reference 17 for maximum emission of noxious products at different operating conditions are also listed. As shown in the table, limits for carbon monoxide (CO) and unburned hydrocarbon (UHC) were specified for engine idle. Limits on smoke and oxides of nitrogen (NO_x) were stated for the takeoff power setting. The emissions of CO, UHC, and NO_x are specified in grams per kilogram of fuel burned for engines fueled with Jet A. NO_x emission from the LH_2 engine is stated in terms of grams per kilogram of fuel burned divided by 2.8 to reflect fuel flow rate for approximately equal thrust levels from both engines. The 2.8 factor is the ratio of heats of combustion of the two fuels, LH_2 divided by Jet A.

In addition to these emissions, it is interesting to note the difference in two other products of combustion which offer contrast between choices of fuels for commercial transport aircraft: water vapor (H_2O) and odors. Water vapor emission is specified in total pounds produced by the specified aircraft per nautical mile during cruise assuming 100% conversion of H_2 to H_2O. No quantitative measure for odors is given, simply a statement that with Jet A fuel there is the familiar odor of kerosene which is objectionable to most people, whereas no odor results from the combustion of hydrogen and air.

The levels of emissions of CO, UHC, smoke, and NO_x listed in Table 18 for combustion of Jet A fuel in advanced design turbofan engines were obtained from work reported by Pratt & Whitney® Aircraft,[20] General Electric Co.,[21] and NASA-Lewis Research Center.[22,23] These sources conclude that the values listed as goals for CO and UHC at engine idle probably will not be met. As previously noted herein, use of LH_2 eliminates concern for CO, UHC, and smoke, and there is strong likelihood that NO_x emissions can be significantly reduced below that forecast for an engine burning Jet A fuel, based on equivalent technology level.

Water vapor from the hydrogen fueled airplane is shown to be nearly twice the quantity produced by the Jet A design. It is interesting to note that the quantity of water shown, which was calculated for the Mach 0.85 cruise, 5500-nmi range, 400-passenger LH_2 aircraft, would be a ribbon only 0.00008 in. thick if confined to four strips each equal in width to the diameter of the engine exhaust nozzles. Compared with the quantity of water vapor normally present in the atmosphere, this does not seem to be a significant addition.

2.4.2.4. Aircraft Performance

A summary of characteristics for a complete spectrum of payload-range combinations of Mach 0.85 subsonic transport aircraft is presented in Tables 19 to 21. Each of these tables lists data for the preferred design of LH_2-fueled aircraft for performing the stated payload-range mission, presents the same data for a corresponding design of Jet A-fueled vehicle, and then lists a ratio which compares the Jet A design with the LH_2. The only exception to this presentation of comparative data is in the case of the 600- and 800-passenger aircraft. No Jet A fueled competitive aircraft designs were generated for those passenger loads.

Figure 23 is a graphical presentation of the amount of energy required to transport passengers in aircraft designed to cruise at Mach 0.85 using both fuels. The energy required to perform various payload-range missions is displayed as a function of the mission capability (expressed in seat nautical miles). Two characteristics are reflected in the plot: the trend of energy requirement for aircraft of a given passenger capacity — with range as the variable; and the energy requirement of aircraft designed for a given range — with passenger capacity as the variable.

The figure shows that the energy requirement varies almost linearly as passenger capacity increases from 400 to 800 seats in aircraft designed for a given range. On the other hand, as the range requirement changes in aircraft designed for a constant number of passengers, the energy requirement varies exponentially. In other words, more energy is required to increase the mission capability (seat nautical mile) of a given aircraft configuration by increasing its range than by adding to its passenger seating capacity. It is also apparent that the energy requirement for Jet A-fueled aircraft increases substantially faster than for aircraft fueled with LH_2.

Figure 24 shows the variation of gross weight, energy utilization, and the difference in energy required to fly the design mission for Mach 0.85 subsonic transport aircraft as a function of range. Both Jet A- and LH_2-fueled designs are represented so the difference in growth trends can be observed. The growth required by the Jet A aircraft as range increases is much more rapid than for the LH_2 designs.

The difference in energy required by aircraft fueled with Jet A compared with those designed to use LH_2 is shown in the upper part of the figure. All the aircraft represented in this plot are designed to carry 400 passengers. Extrapolation of the curve indicates that the energy advantage of the LH_2-fueled aircraft for this passenger capacity approaches zero at a design range of slightly less than about 2000 nmi. However, it is also evident that there will be no significant disadvantage for the LH_2 aircraft, even at very short ranges.

TABLE 19

Summary of Characteristics — Short Range, Mach 0.85 Passenger Aircraft

		130 PAX — 1500 nmi		
	Units	Jet A	LH$_2$	Ratio (Jet A/LH$_2$)
Weight, gross	lb	108,700	98,260	1.11
Weight, total fuel	lb	19,700	7,360	2.68
Weight, operating empty	lb	60,350	62,290	0.97
Thrust per engine (SLS)	lb	18,910	16,950	1.12
Wing area	ft^2	928.7	911.5	1.02
Span	ft	101.1	95.5	1.06
Fuselage length	ft	113.0	139.5	0.81
Fuselage diameter	ft	13.0	13.5	0.96
L/D, cruise	—	16.3	13.9	1.18
SFC, cruise	(lb/hr)/lb	0.618	0.211	2.93
Federal air regulation takeoff distance	ft	7970	7885	1.02
FAR landing distance	ft	5754	5728	1.00
Cost, Research, development, test, and engineering	$10^6	23.7	21.6	1.10
Cost, production	$10^6	7.51	7.85	0.96
Cost, Δ cost for LH$_2$[a]	$/10^6 Btu		0	
Noise, flyover	EPNdB	79	79	—
Noise, sideline	EPNdB	86	86	—
Noise, approach	EPNdB	90	91	—
Energy utilization	Btu/seat nmi	1288	1339	0.96

[a]   Extra cost which can be paid for LH$_2$ to yield equal DOC; based on 28.6¢/gal for Jet A (cost to U.S. domestic carriers in September 1975).

TABLE 20

Summary of Characteristics — Medium Range, Mach 0.85 Passenger Aircraft

Payload-Range	Units	200 PAX — 3000 nmi			400 PAX — 3000 nmi			3000 nmi	
		Jet A	LH_2	Ratio (Jet A/LH_2)	Jet A	LH_2	Ratio (Jet A/LH_2)	600 PAX LH_2	800 PAX LH_2
Weight, gross	lb	216,900	179,500	1.21	404,300	335,200	1.21	509,800	697,500
Weight, total fuel	lb	61,140	20,920	2.92	105,700	34,300	3.08	49,130	63,700
Weight, operating empty	lb	111,800	114,500	0.98	210,600	212,900	0.99	328,600	457,800
Thrust per engine (SLS)	lb	15,290	15,030	1.02	25,770	24,720	1.04	35,050	46,200
Wing area	ft^2	1664	1602	1.04	3235	3047	1.06	4655	6400
Span	ft	127.4	123.4	1.03	170.6	165.6	1.03	204.7	240.2
Fuselage length	ft	144.7	173.4	0.83	197	210	0.94	236.0	266.7
Fuselage diameter	ft	19.58	19.58	1.00	19.26	21.75	0.89	24.67	26.33
L/D, cruise	—	15.3	13.8	1.11	16.7	14.9	1.12	15.9	16.9
SFC, cruise	lb hr/lb	0.616	0.211	2.92	0.582	0.200	2.91	0.200	0.200
FAR takeoff distance	ft	7975	5380	1.48	7980	5860	1.36	6188	6345
FAR landing distance	ft	5760	5780	1.00	5760	5804	0.99	5823	5833
Cost, RDT&E	$10^6	391	362	1.08	917	1002	0.92	1355	1715
Cost, production	$10^6	13.3	14	0.95	22.6	23.4	0.97	34.0	45.0
Cost, Δ cost for LH_2[a]	$/10^6 Btu		0.14			0.22			
Noise, flyover	EPNdB	86	82	—	93	88	—		
Noise, sideline	EPNdB	86	86	—	86	86	—		
Noise, approach	EPNd	93	93	—	97	98	—		
Energy utilization	Btu seat nmi	1537	1464	1.05	1260	1204	1.05	1146	1112

[a] Extra cost which can be paid for LH_2 to yield equal DOC; based on 28.6¢/gal for Jet A (cost to U.S. domestic carriers in September 1975).

TABLE 21

Summary of Characteristics — Long Range, Mach 0.85 Passenger Aircraft

Payload-Range	Units	400 PAX — 5500 nmi			5500 nmi		400 PAX — 5000 nmi radius		
		Jet A	LH_2	Ratio (Jet A/LH_2)	600 PAX LH_2	800 PAX LH_2	Jet A	LH_2	Ratio (Jet A/LH_2)
Weight, gross	lb	523,200	391,700	1.34	614,500	840,000	992,500	587,400	1.69
Weight, total fuel	lb	190,800	61,600	3.10	93,320	122,400	524,000	150,800	3.47
Weight, operating empty	lb	244,400	242,100	1.01	389,100	541,600	380,500	348,500	1.09
Thrust per engine (SLS)	lb	32,700	28,700	1.14	41,480	53,550	49,630	39,350	1.26
Wing area	ft^2	4186	3363	1.24	5297	7304	7125	5020	1.42
Span	ft	194.1	174	1.12	218.3	256.4	280	224	1.25
Fuselage length	ft	197	219	0.90	268.5	300.2	225	254	0.89
Fuselage diameter	ft	19.26	21.75	0.89	24.67	26.33	19.58	21.75	0.90
L/D, cruise	—	17.9	16.1	1.12	16.7	17.5	20.3	16.8	1.21
SFC, cruise	lb_{hr}/lb	0.581	0.199	2.92	0.199	0.200	0.533	0.199	2.93
FAR takeoff distance	ft	7990	6240	1.28	6686	7000	11,973	6914	1.73
FAR landing distance	ft	5210	5810	0.90	5831	5830	5867	5890	1.0
Cost, RDT&E	10^6	1109	1124	0.99	1567	1960	1222	920	1.33
Cost, production	10^6	26.6	26.9	0.99	40.9	54.0	40.0	38.9	1.03
Cost, Δ cost for LH_2[a]	$/$10^6$ Btu		0.44		—	—		1.05	
Noise, flyover	EPNdB	94	89	—			100	93	—
Noise, sideline	EPNdB	88	87	—			93	94	—
Noise, approach	EPNdB	97	98	—			96	98	—
Energy Utilization	Btu / Seat nmi	1384	1239	1.12	1251	1226	2122	1695	1.25

[a] Extra cost which can be paid for LH_2 to yield equal DOC; based on 36.6¢/gal for Jet A (cost to U.S. international carriers in September 1975).

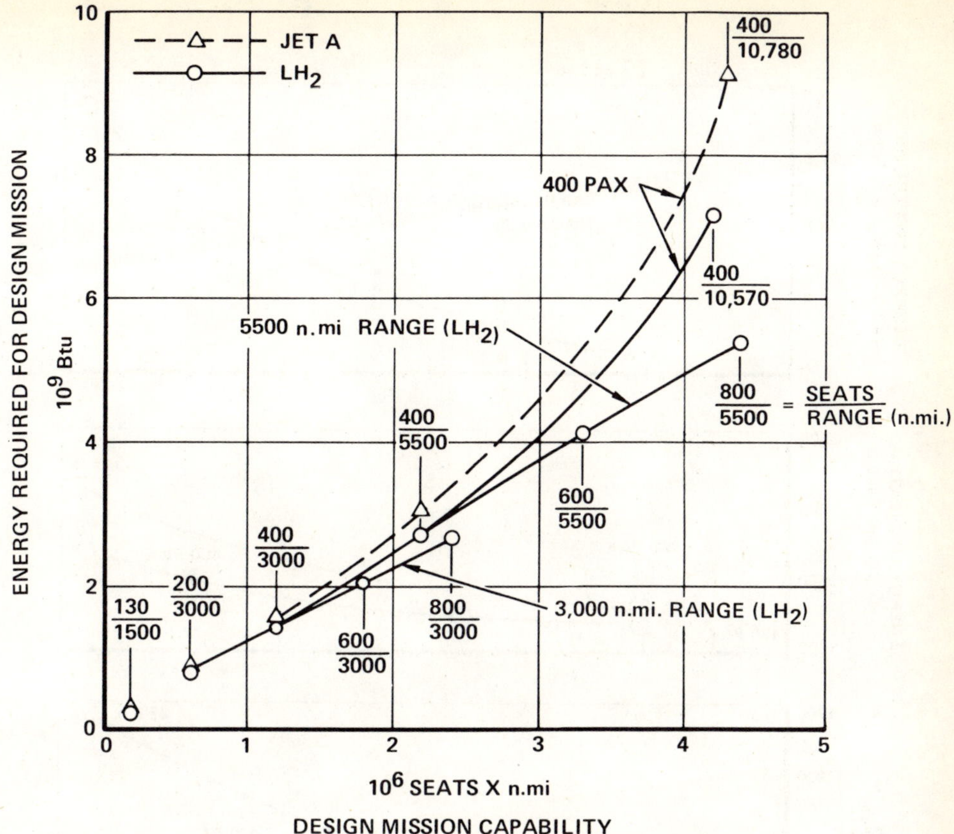

FIGURE 23.   Energy vs. design mission capability for Mach 0.85 aircraft.

In summary, the advantage of using LH_2 as fuel in subsonic transport aircraft generally increases with the amount of energy required to perform the mission. The crossover point, above which LH_2 can be used to advantage and below which Jet A is slightly more energy efficient, varies with the passenger load. For the 130-passenger Mach 0.85 aircraft shown in the lower left corner of Figure 23, the crossover point is approximately the 1500-nmi. design range, which requires about 0.25×10^9 Btu. For a 400-passenger Mach 0.85 aircraft, the crossover appears to be just under the 2000-nmi. design range, a mission which needs approximately 10^9 Btu.

2.5. SAFETY

No discussion of the possible use of hydrogen as fuel, particularly in aircraft, can be complete without some mention of safety. There is a widespread conception among the general public that hydrogen is a highly unstable material, unusually susceptible to deflagration and detonation. The facts do not support this view.

Nevertheless, mention hydrogen as a fuel for passenger-carrying aircraft, and for most people an immediate response is a mental image of the German airship *Hindenburg* burning near its mooring mast at Lakehurst, N.J. in May 1937. This accident, tragic as it was, has proved to be far more serious as a psychological deterrent to further widespread use of hydrogen than can possibly be logically justified.

In the *Hindenburg* disaster, 35 people out of a total of 97 who were on board lost their lives. In the present time period, when more than 1000 people are killed every

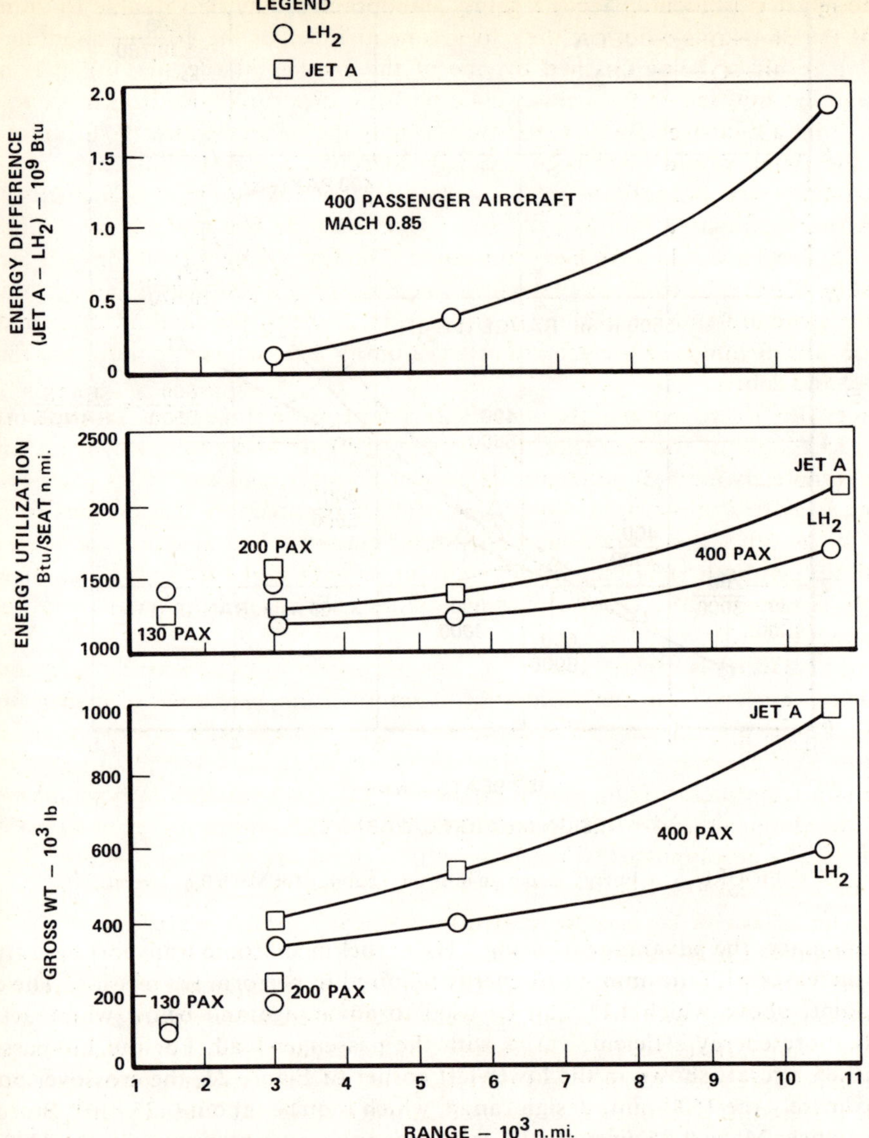

FIGURE 24.   Growth characteristics of subsonic aircraft vs. range.

week in the U.S. in motor vehicle - related accidents, one wonders why this single event has been accorded such significance in the public awareness. The probable answers are that:

1. The *Hindenburg* was the first commercial airship disaster.
2. It was photographed by newsreel cameras and, because it was so spectacular, the films were shown around the world.
3. A routine radio broadcast describing the airship docking has become a part of radio legend because of the announcer's hysterical reaction as the craft burst into flames.
4. The *Hindenburg,* filled with gaseous hydrogen, is the passenger-carrying airship which is most often recalled to account for the passing of a concept.

Although it is academic, from a safety standpoint it may be valuable to point out that, of the 36 people who lost their lives (one member of the ground handling crew died as a result of being crushed by one of the engines of the airship), it is highly probable that not one of the passengers died from exposure to hydrogen flames. Radiation from a hydrogen flame is so low that only the structure directly adjacent to or above the flame would have been seriously affected. Since the passenger and crew compartments were beneath the gas bags, the people should not have been burned.

It has been estimated that less than a minute after the fire was initiated, all of the gaseous hydrogen would have been consumed. The fire which burned for over an hour was fed by diesel oil carried as fuel for the engines, plus wood, fabric, aluminum, etc. from the structure and furnishings of the airship. Most of the fatalities occurred as a result of falls or jumps to the ground or from burns suffered as a result of the flaming diesel oil and debris.

People who are concerned about the safety aspects of riding in a hydrogen-fueled aircraft lose sight of the fact that the mixture of liquid and gaseous hydrogen in the tank is completely inert. It needs air, or oxygen, to react, and air cannot enter because the tank is always pressurized. Hydrogen is a fire hazard only if it escapes from the tank. On the other hand, gasoline tanks in automobiles and kerosene tanks in commercial aircraft both characteristically contain a mixture of fuel vapor and air which is highly flammable and, in the proper mixture ratio, explosive. All it requires for ignition is a spark.

On an overall basis, most experts consider hydrogen to be at least as acceptable as gasoline or kerosene from the standpoint of safety. Some properties of hydrogen tend to make it more hazardous than hydrocarbon fuels, while other properties make it less hazardous (see Chapter 5 for a complete discussion of safety). The remarkable record of freedom from incidents involving liquid and gaseous hydrogen in both the experimental development and operational phases of the U.S. space program is testimony to what can be accomplished when proper procedures are followed. Indeed, the long period of use of gaseous hydrogen as a lifting agent for the German Zeppelins is proof that the material can be used successfully in an application fraught with far more hazards than the use contemplated here.

Table 22 presents a tabulation of several events which can occur in manufacturing, transporting, storing, and/or handling fuel for commercial transport aircraft and assesses the comparative risk from hydrogen and kerosene (Jet A). It is clear that operational procedures for storing and handling fuel for commercial transport aircraft must be different if hydrogen is to be used in place of Jet A. The fact that the equipment and procedures are different should not be allowed to deter implementation and use of the superior fuel system. The excellent safety record attained in the U.S. in the last 20 years when millions of gallons of LH_2 have been used is evidence that the procedures which have been established are practical and effective.

2.6. SCENARIO FOR IMPLEMENTATION

The results reported in the foregoing sections clearly show that LH_2 is a very attractive fuel for both supersonic and subsonic transport aircraft. This leads to consideration of what must be accomplished to implement its use in such aircraft in the future.

The following scenario is presented as a feasible occurrence, not as a prediction of what might actually happen. There are too many major uncertainties involved in the sequence of events which must take place before a happening as significant as the contemplated change of fuel system for commercial aviation can occur to warrant suggesting a high probability for the timetable outlined. The actual events which take

TABLE 22

Comparison of Hydrogen and Kerosene on Basis of Safety Aspects

Event	Effect		Considered to be an advantage for
	Hydrogen	Kerosene	
A spill or leak during bulk handling	LH_2 is heavier than air, but it would immediately vaporize; the gaseous H_2 would dissipate by rising rapidly; no contamination of surroundings	Kerosene and its vapor are both heavier than air and would, therefore, flow and collect in low areas; serious contamination of surrounding area	Hydrogen
Flammability hazard	Energy level required for ignition is low, but ignition temperature is relatively high (1085°F); flammability limit in air is 4.1 to 74% by volume; it burns very rapidly with flame of low emissivity (0.01 to 0.10) which minimizes radiation effects; no smoke	Ignition temperature is quite low (500°F); flammability limit in air is 1.0 to 7.6% by volume; a comparable quantity would burn 10 to 15 times as long; emissivity (1.0) is 10 to 100 times that of H_2 with consequent increase in potential for damage to surrounding structure; copious smoke, CO, and other noxious products	Hydrogen
Detonation hazard			
Unconfined	Low	Low	Neither
Confined	High	Low	Kerosene
Human exposure	Small quantities of LH_2 spall on contact with skin and immediately vaporize with no damage; larger quantities freeze tissue with results similar to severe burns; vapors are nontoxic	Liquid kerosene is a minor skin irritant; vapors are toxic at levels of 500 ppm	Hydrogen
Temperature increase from normal storage condition	LH_2 boils at −423°F at atmospheric pressure; its vapor pressure rises rapidly with an increase in temperature resulting in a loss from venting, rupture of a tank, or a requirement that the liquid be rerefrigerated	Small increase in volume of the liquid	Kerosene
Contamination of and by the fuel	Problem does not exist with LH_2 because of the extremely low temperature; LH_2 is essentially a pure substance after it is liquified, and it is too cold to support organic growth; it is noncorrosive but does cause embrittlement of some metals	Hydrocarbon fuels have serious problems with both organic and inorganic contamination and with corrosion of some metals	Hydrogen

place before this change is made are dependent on major uncertainties such as the following:

1. An authoritative decision being made to have the commercial air transport industry become an early, major user of hydrogen as fuel for new, advanced design aircraft
2. The timing and priority assigned to this decision
3. The efficacy with which a plan is implemented to create plants to manufacture hydrogen in sufficient quantities and for designated airports to be equipped with necessary liquefaction, storage, and handling facilities.
4. Coordination of U.S. emphasis on aircraft usage of LH_2 with that of other countries which are major participants in international air travel.

Considering the serious nature of the problems associated with assuring an adequate world-wide supply of petroleum-based fuel at an economically acceptable price after about 1990 and in view of the many attractive advantages which accrue from using LH_2 as fuel in advanced designs of commercial transport aircraft, it is felt that the possibility of necessary positive action being taken is high and that the timetable suggested in Figure 25 for implementing this change is feasible.

It should be recognized, and is hereby emphasized, that the need is great for a rigorous analysis of all the interrelationships involved in the problem of changing fuel systems for the air transport industry. A comprehensive societal impact study should be made to explore properly the ramifications such a change would make in established economic, industrial, commercial, regulatory and social processes.

Figure 25 presents the highlights of developmental activity which must precede operation of LH_2-fueled commercial transport aircraft between an initial city-pair whose

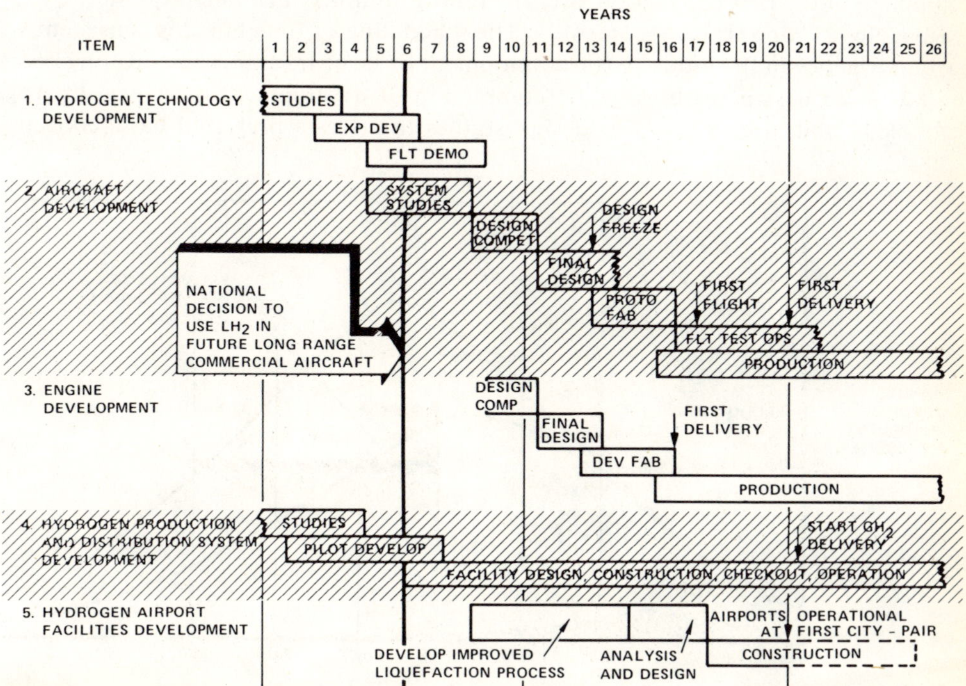

FIGURE 25.   Timetable of events leading to operational usage of LH_2 transport.

airports are suitably equipped. Continued development and expansion of airport facilities within the U.S. and internationally can lead to operation of significant numbers of long-range, LH₂-fueled transport aircraft within 5 years following the initial operation (Figure 26).

The development activity shown in the figure is divided into five major categories. Each of these elements of the scenario must be addressed and successfully accomplished in order for the end objective to be achieved. It is indicated that within 6 years properly directed studies and development could provide a basis for a national decision to use hydrogen fuel in transport aircraft of the future. Within 15 years from that decision date the first commercial operations could become a reality if the project is given adequate priority.

Highlights of the five categories of development activity are as follows.

Item No. 1: Hydrogen technology development — The first category is development of hydrogen technology for aircraft application. Some examples of the types of items which would be included in this activity are (1) aircraft design studies to discover preferred configurations for LH₂-fueled aircraft of various payload-range capabilities; (2) tank and insulation system design and development; (3) aircraft fuel system design and development with special emphasis on boost pumps, transfer lines, and valves; (4) advanced aircraft engine design studies and development of those components whose characteristics will be peculiar to use of hydrogen, e.g., heat exchangers, combustors, and high-pressure pumps; and (5) design and development of critical elements of other aircraft fuel subsystems, such as fueling and defueling, venting and pressurization, and fuel quantity and flowrate measurement devices which can meet aircraft operational and service life requirements. A program of technology design studies has already been initiated by NASA. It should be actively pursued in order to provide the special knowledge of hydrogen-peculiar equipment and systems needed to complete design and development of the first production aircraft, Item 2, in timely fashion.

Item No. 2: Aircraft development — The scheduling of Item No. 2 is consistent with current practice in the industry for development of large transport aircraft incorporating advanced design features. After completion of development of critical hydrogen technology and after a series of design studies to select a preferred basic concept, 2

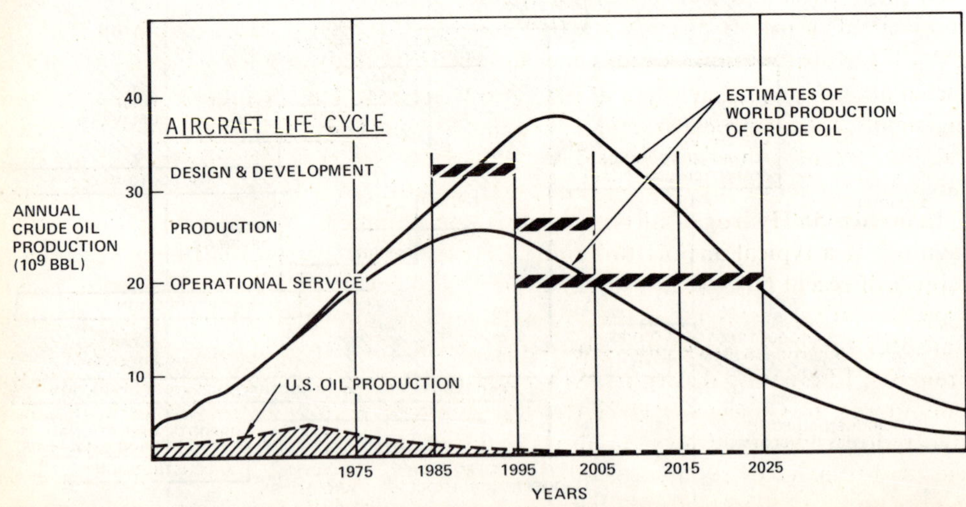

FIGURE 26.   Correlation of life cycle of new transport aircraft with forecasts of oil availability.

years is permitted to establish detail design of the production aircraft. After design freeze and while final design details are completed, fabrication of long lead time items is begun. Fabrication of the first aircraft can be completed in just over 3 years, 6 years after selection of a preferred design concept. First flight of this aircraft could occur approximately 1 year later after a program of extensive ground testing. Delivery of the first aircraft for operational airline service would normally follow about 3 years later. Initial commercial operation of a hydrogen-fueled transport aircraft could conceivably occur in 1995 if the crucial national decision is made in 1980. Normal build-up of production deliveries would result in 22 aircraft being put in service the first year, 48 the second year, and 220 within 5 years. The build-up of production of the aircraft can be much faster than deliveries can be assimilated in commercial operations. Development of gaseous hydrogen production capability (Item No. 4) and airport facilities (Item No. 5) will pace the growth of LH_2 transport aircraft usage.

Item No. 3: Engine development — Engine development would proceed in parallel with the aircraft development so delivery of the first set of engines for installation on the prototype aircraft could occur approximately 1 year before first flight.

Item No. 4: Hydrogen production and distribution system development — Development of a capability for production and distribution of adequate quantities of gaseous hydrogen (GH_2) will require immediate priority attention if initial commercial operation of LH_2-fueled aircraft is to occur in 1995. This is the critical pacing item of the entire undertaking. The quantities of GH_2 required to support airline usage of long-range, wide-bodied aircraft in the time period starting in the 1990 decade will require dependence on production processes which are currently understood and basically developed, aside from steam reforming of natural gas or partial oxidation of crude oil, for which neither resource can logically be considered to be available for the proposed purpose. The production processes which are leading candidates involve gasification of coal and organic wastes and electrolysis of water. All of these processes will require long lead times for development of a capability for supplying adequate quantities of GH_2. The lead time for building new nuclear reactors is currently 10 to 12 years. The time required to expand our coal mining capability adequately will be nearly as long. Clearly, it will take a high order of incentive and resolve, similar to that demonstrated in the Manhattan Project and in the U.S. Apollo "Man on the Moon in this Decade" program, to accomplish the tasks required to have adequate GH_2 production and transmission capability available in time to supply the needs of commercial transport aircraft starting early in the 1990 decade, assuming that "go-ahead" for the program to commence development is given in 1980. Initial use of LH_2-fueled aircraft can occur when at least two airports which constitute a city-pair involving significant reciprocal traffic are equipped with LH_2 fueling and maintenance capability. Realistically, it is considered that 1995 is a credible date to indicate initial capability for supplying GH_2 in substantial quantities for liquefaction at two airports.

Item No. 5: Hydrogen airport facilities development — The analysis of facilities required at a typical airport to handle LH_2 fueled aircraft in the future has been the subject of recent NASA-sponsored studies.[24,25] The studies were performed to provide a preliminary assessment of the problems and requirements of handling LH_2 fueled transport aircraft at a designated airport. Lockheed-California Company[24] studied San Francisco International Airport (SFO) and the Boeing Company[25] analyzed the problems unique to Chicago O'Hare (ORD). A major part of the effort in both studies involved consideration of the facilities required at these airports for liquefaction, storage, and transfer of the hydrogen to aircraft. The resulting estimates are that it will require about 42 months for design and construction of the first 250 ton/day hydrogen liquefaction plant. Succeeding plants can be expected to be built in 36 months. Four such plants would be required to support anticipated traffic at both SFO and ORD in 1995. Construction of other facilities required at the airport to permit servicing LH_2-fueled aircraft can be accomplished within this time frame. Accordingly, it is felt that development of hydrogen liquefaction, storage, and handling facilities at selected airports around the country, with proper lead time and planning, can proceed on a schedule which matches the projected availability of the GH_2 from Item No. 4.

2.7. CONCLUSIONS

Despite the obvious advantages of using LH_2 as fuel in commercial transport aircraft, there will be considerable resistance to changing from Jet A. Three arguments are heard most often; first, that hydrogen is uneconomical. LH_2 will, indeed, cost more to produce from coal than does synthetic Jet A. However, the lower energy requirement of LH_2-fueled vehicles in performing the longer range missions compensates for the difference.

Witcofski[26] presents data which show, for example, that when coal sells for about \$17.50/ton, hydrogen can be produced in a modern steam-iron process and can be liquefied using current technology for \$6 per million Btu. Alternatively, the coal can be liquefied using the Consol® Synthetic Fuel process and then refined by hydrocracking and hydrogenation to make synthetic Jet A (synjet) for approximately $\$5/10^6$ Btu. The overall efficiencies to produce these liquid fuels from coal are 49 and 54% respectively.

Based on these process efficiencies and considering the Mach 2.7 supersonic aircraft whose characteristics are presented in Table 9, the Jet A version would require 470 tons of coal to produce the fuel needed to fly the design mission. The LH_2 SST would need only 375 tons of coal, a net saving of over 25%.

At the fuel prices quoted, the coal-derived synthetic jet fuel required to transport 234 passengers 4200 nmi at Mach 2.7 would cost \$30,400. The coal-derived LH_2 would cost only \$26,400, a saving of 15%.

In the case of the 400-passenger, 5500-nmi range subsonic transport aircraft described in Table 15, the results are as follows:

	Synjet-fueled aircraft	LH_2-fueled aircraft
Tons of coal required to produce fuel for design mission	235	232
Cost of coal-derived fuel for design mission	\$15,226	\$16,375

Thus, on the basis of a preliminary evaluation of the energy and cost of producing LH_2 and for an early evaluation of the performance potential of LH_2-fueled aircraft designs, supersonic vehicles show a significant cost and energy advantage, and subsonic designs are closely competitive with equivalent synjet-fueled models.

As noted, the liquefaction process used as a basis for estimating the energy and cost of LH_2 in the example from Witcofski[26] represents current technology. A study underway for NASA by the Linde Division of Union Carbide Corporation[27] has already identified a 20% improvement potential in the energy requirement. Obviously, this will result in a decreased cost for the LH_2 operation.

Similarly, preliminary results from a study of fuel systems for LH_2 fueled subsonic transport aircraft, Contract NAS 1-14614 being performed by Lockheed-California

Company for NASA-Langley Research Center,[19] indicate that the energy requirement of the synjet-fueled aircraft in performing the design mission is more like 20% greater than that of the LH_2 design, rather than 12% as listed in Table 15, herein. When final results of these studies are incorporated in an analysis of relative energy and costs, the LH_2-fueled aircraft can be expected to show significantly greater potential in both the supersonic and subsonic speed regimes, in addition to the performance and environmental advantages which are obtained.

An even more fundamental counter to the cost/energy argument, however, is the fact that commercial aircraft must be capable of operating world-wide. Since coal and oil shale are not uniformly available as natural resources around the world, there are many areas to which either syncrude or the synthetic Jet A would have to be shipped. It is not reasonable to assume that 25 years from now the U.S.S.R. and the U.S., which together possess over 75% of the known coal reserves in the world, will permit export of large quantities of this nonreplenishable resource which is obtained at the cost of digging up tremendous areas of the countryside. Current experience in the U.S. indicates that environmental controls will probably limit coal or shale production to the amount needed for domestic consumption, and even that will undoubtedly be subject to severe restrictions.

On the other hand, if hydrogen is the fuel used in commercial transport aircraft, every country in the world with access to water can manufacture its own fuel, using whatever source of energy is most convenient — nuclear, solar, geothermal, hydro, or organic waste products — in addition to using any fossil resource which may be locally available. Then, neither economic nor diplomatic sanction can restrict aircraft fuel supplies, and the price structure should be stable. On a world-wide basis, it seems clear that the most equitable choice of fuel for future use in commercial transport aircraft is LH_2.

The second argument frequently heard in opposition to early use of hydrogen in aircraft is the huge sum of money required to provide the logistic structure needed for the new cryogenic fuel. There is no denying the enormous cost. The point is that, since the supply of petroleum in the world is finite, the change must be made ultimately; the only question is when.

When production of Jet A from petroleum no longer is economically acceptable, the air transport industry will turn to one of the three alternate fuels: synthetic Jet A, liquid methane, or liquid hydrogen, each of which can be made from coal. There will be strong incentive within the U.S. to adopt synthetic Jet A. There is adequate coal and oil shale to supply domestic needs of aircraft plus other needs, for probably two centuries. With proper processing, the synthetic product can be used as a direct replacement for petroleum-based Jet A so no change in either aircraft or airport handling equipment will be required. Obviously, this is a significant advantage.

However, when the international requirements for aviation jet fuel are considered, it clouds this picture of synthetic Jet A being an ideal alternative. In fact, synthetic Jet A would be a poor choice as a replacement fuel for many nations of the world because neither coal nor oil shale are available in sufficient quantities.

Liquid hydrogen rather clearly should be the preferred choice of alternate fuel for both long- and short-range transport aircraft around the world because of its performance potential, its environmental advantages, and the fact that it can be manufactured from water using any of a large number of energy sources, at least one of which is certain to be available anywhere in the world.

The change to an alternate fuel must be made before the capability of supplying the needs of the world with petroleum deteriorates to the point that it triggers international economic chaos. Given the long lead time required to develop manufacturing facilities to make hydrogen for long haul commercial aircraft, plus consideration of the 35- to 40-year design/production/use cycle of transport aircraft, it is clear that the time for serious consideration leading to a decision is upon us. Figure 26 illustrates the point.

The characteristic life cycle of large transport aircraft of advanced design is shown superimposed on a plot of two forecasts of the future of crude oil production in the world. The lower curve of oil production is based on a conservative estimate that a total of 1350×10^9 barrels will ultimately be produced. The upper curve reflects a more optimistic projection of 2100×10^9 barrels. The significant point to note is that the peak of production from both of these estimates by responsible groups of scientists occurs within a 10-year span, the latest being the year 2000. If a new aircraft design is started in 1985, by the time the initial production airplanes are placed in operational service, the price and availability of petroleum-based Jet A will have begun to be influenced by the impending shortage of the resource.

The case for an early decision to use LH_2 fuel in future designs of long-range aircraft is much stronger than the argument for delaying the decision, thereby risking a future demand for fossil fuels that exceeds the capability to produce them. As stated, the cost of changing fuels will be great. However, commercial air operations are expected to grow approximately 5% per year during the next 25 years. At that rate, if the decision to change fuels is put off until 2000, the industry will be three times as big, and the trauma and the cost of converting to hydrogen will be at least three times as great.

Finally, the third argument usually advanced is that commercial aircraft do not use much fuel on a percentage basis, and the "others" should switch to an alternate so aircraft can be provided with conventional Jet A fuel well into the 21st century. There are two aspects of this argument which require comment. The first is that the chemical composition of a barrel of crude oil dictates the amount of any given product that can be processed from it. The amount of aviation grade kerosene currently derived in the U.S. from a barrel of crude oil is only 5% of the product output. Refineries can be reprogrammed to increase production of Jet A for commercial aviation, to approximately 17% of the barrel. However, this would revise the output of the other products so that production of gasoline for motor fuel, for example, would drop from 47 to about 18% and diesel fuel would drop from 26 to 20%. In short, maximizing production of Jet A is not a practical answer to the aircraft fuel supply problem unless an alternate fuel is developed for the largest user of petroleum products: the automobile, truck, and bus market. To date, no strong candidate has emerged that would permit consideration of this possibility.

The second aspect of this question of "let someone else switch" is the scope of the task. At the present time, commercial transport aircraft use only about 4.1% of the petroleum consumed in the U.S. The magnitude of the undertaking to manufacture sufficient quantities of LH_2 from coal and/or by electrolysis of water to meet the requirements of commercial aviation is dwarfed by the size of the effort which would be needed to provide an alternate fuel for the larger requirements of the truck, automobile, and bus market or for the industrial and electric utility markets, the other large-scale users of petroleum. To address this larger market first would protract the period during which the change to the new fuel is accomplished and would compound the problems. Simple logic dictates the introduction of an alternate fuel in an industry where the demand is small enough that it can be met with a more manageable effort to provide the new infrastructure required.

In summary, the end is in sight for the world supply of petroleum. Serious consideration must be given to alternate sources of energy and fuels which can be used when petroleum is economically unavailable. The decisions concerning which fuels should be developed for which purpose and for the timing of their initiation should be reached after careful deliberation. For industries like air transport, where major equipment is expected to be in service 35 to 40 years after start of design, the need for an early decision concerning designation of a fuel of the future is most urgent. Lack of a decision or, for example, delay after 1980 can have significant long range cost implications.

Liquid hydrogen is recommended as a logical possibility for fuel in future designs of long-range commercial transport aircraft and as a candidate for early changeover.

REFERENCES

1. **Hubbert, M. K.,** The energy resources of the earth, *Sci. Am.,* 224 (3), 149, 1971.
2. **Skinner, B. J. and Emery, K. O.,** *Mineral Resources and the Environment,* National Academy of Sciences, Washington, D.C., 1975.
3. **Hoehling, A. A.,** *Who Destroyed the Hindenberg?,* Popular Library, New York, 1962.
4. **Silverstein, A. and Hall, E. W.,** Liquid Hydrogen as a Jet Fuel for High-Altitude Aircraft, NACA RM E55C28a, National Advisory Committee for Aeronatics — Lewis Flight Propulsion Laboratory, Cleveland, Ohio, 1955.
5. **Fenn, D. B., Acker, L. W., and Algranti, J. S.,** Flight Operation of a Pump-Fed, Liquid-Hydrogen Fuel System, NASA TM X-252, NASA-Lewis Research Center, Cleveland, Ohio, April 1960.
6. **Mulready, R. C.,** Liquid hydrogen engines, in *Technology and Uses of Liquid Hydrogen,* Scott, R. B., Denton, W. H., and Nicholls, C. M., Eds., Macmillan, New York, 1964, Chap. 5.
7. **Edeskuty, F. J.,** Liquid hydrogen as a coolant/propellant for nuclear rockets, in *Technology and Uses of Liquid Hydrogen,* Scott, R. B., Denton, W. H., and Nicholls, C. M., Eds., Macmillan, New York, 1964, Chap. 6.
8. **Brewer, G. D.,** Manned hypersonic vehicle, in *Lockheed Horisons,* Issue No. 4, Lockheed-California Company, Burbank, First Quarter 1966.
9. **Peterson, R. H. and Waters, M. H.,** Hypersonic transports — economic and environmental effects, paper presented at ICAS/AIAA Meeting, Amsterdam, September 1972.
10. **Becker, J. V. and Kirkham, F. S.,** Hypersonic aircraft — 21st century transports?, *ICAO Bull.,* p. 8, February 1974.
11. **Morris, R. E. and Williams, N. B.,** Study of Airbreathing Launch Vehicles with Cruise Capability, NASA CR-73194-73199, Volumes 1 to 6, Lockheed-California Co. for NASA-Ames Research Center, Moffett Field, Cal., February to April 1968.
12. **Anon.,** *Standard Method of Estimating Comparative Direct Operating Cost of Turbine Powered Transport Airplanes,* Air Transport Association of America, Washington, D.C., December 1967.
13. **Brewer, G. D. and Morris, R. E.,** Minimum Energy Liquid Hydrogen Supersonic Cruise Vehicle Study, NASA CR 137776, Lockheed-California Company (October 1975).
14. **Johnston, H.,** Reduction of stratospheric ozone by nitrogen oxide catalysts from supersonic transport exhaust, *Science,* 173 (3996), 517, 1971.
15. **Newell, R. E.,** The global circulation of atmospheric pollutants, *Sci. Am.,* 244(1), 32, 1971.
16. **Manabe, S. and Wetherald, R. T.,** Thermal equilibrium of the atmosphere with a given distribution of relative humidity, *J. Atmos. Sci.,* 24(3), 241, 1967.
17. **Brewer, G. D., Morris, R. E., Lange, R. H., and Moore, J. W.,** Study of the Application of Hydrogen Fuel to Long-Range Subsonic Transport Aircraft, NASA CR 132559, Lockheed-California Company and Lockheed-Georgia Company under Contract NAS 1-12972, NASA-Langley Research Center, Hampton, Va., January 1975.
18. **Brewer, G. D. and Morris, R. E.,** Study of LH_2-Fueled Subsonic Passenger Transport Aircraft, NASA CR-144936, Lockheed-California Company, NASA-Langley Research Center, Hampton, Va., January 1976.
19. Study of Fuel Systems for LH_2-Fueled Subsonic Transport Aircraft, Contract NAS 1-14614, Lockheed-California Company, NASA-Langley Research Center, Hampton, Va., work in progress.
20. **Brines, G. L.,** Studies for Determining the Optimum Propulsion System Characteristics for Use in a Long Range Transport Aircraft, NASA CR-120950, NASA-Lewis Research Center, Cleveland, Ohio, 1972.
21. **Anon.,** Propulsion System Studies for an Advanced High Subsonic Long Range Jet Commercial Transport Aircraft, NASA CR-121016, NASA-Lewis Research Center, Cleveland, Ohio, 1972.

22. **Grobman, J., Norgren, C., and Anderson, D.,** Turbojet Emissions, Hydrogen versus JP. NASA TM-X-68258, NASA-Lewis Research Center, Cleveland, Ohio, 1973.
23. **Grobman, J. and Ingebo, R. D.,** Jet Engine Exhaust Emissions for Future High Altitude Commercial Aircraft, NASA TM-X-71509, NASA-Lewis Research Center, Cleveland, Ohio, 1974.
24. **Brewer, G. D., Ed.,** LH_2 Airport Requirements Study, NASA CR-2700, Lockheed-California Company, NASA-Langley Research Center, Hampton, Va., March 1976.
25. **Anon.,** Exploratory Study to Determine Integrated Technological Air Transportation System Ground Requirements of Liquid Hydrogen Fueled Subsonic Long-Haul Civil Air Transports, NASA CR-2699, Preliminary Design Department, Boeing Commercial Airplane Company, NASA-Langley Research Center, Hampton, Va., March 1976.
26. **Witcofski, R. D.,** Alternate Aircraft Fuels — Prospects and Operational Implications NASA TM X-74030, NASA-Langley Research Center, Hampton, Va., May 1977.
27. **Anon.,** Survey Study of the Efficiency and Economics of Hydrogen Liquefaction, NASA CR-132631, Linde Division of Union Carbide Corporation, NASA-Langley Research Center, Hampton, Va., April 1975.

Chapter 3

Domestic Uses of Hydrogen

Chapter 3

DOMESTIC USES OF HYDROGEN

J. Pangborn and M. I. Scott

TABLE OF CONTENTS

3.1. INTRODUCTION

The use of hydrogen as a residential fuel can be justified if hydrogen gas can be produced, delivered to, and utilized by residential customers more economically than can other energy forms. Hydrogen is, thus, a potential alternative to natural gas, oil, or electricity. The substitution of hydrogen for natural gas or oil would come about only if the prices or availability of these fuels dictated the change.

It should be noted that total conversion to hydrogen-energy utilization is not being advocated here. It is anticipated that a combination of energy forms will continue to be consumed, i.e., fuel gas and electricity. Some applications, such as lighting require-ments and electric motors, are best served by electricity. However, eventually it may be beneficial to convert applications that currently consume fossil fuel (primarily nat-ural gas) to hydrogen consumption because a distribution network already exists that has utilization equipment potentially capable of using hydrogen as a fuel after some modification. This chapter discusses the existing natural gas distribution system, the compatibility of hydrogen in the distribution system, residential use patterns, conver-sion of existing appliances, and the development of catalytic appliances.

3.2. THE NATURAL GAS DISTRIBUTION SYSTEM

The U.S. gas distribution system typically consists of one or more networks of piping that carry the gas to the ultimate consumers from the various sources of supply: city gate station, gas-storage facilities, and gas-manufacturing plants. Figure 1 illustrates the various components of a representative distribution system that includes a manu-factured gas supply.[2] The latest available statistics show that purchased and produced gas is $18,730 \times 10^{12}$ Btu/year natural gas and 30×10^{12} Btu/year manufactured and substitute natural gas (SNG). Annual production of the latter is growing at a very rapid rate. Gas now constitutes more than one fourth of the U.S. total energy supply. The basic function of the city gate stations is to meter the natural gas and to reduce the gas pressure of the transmission pipeline to that of the distribution system. Most stations measure the gas with orifice meters, although other types of meters are some-times used. Pressure reduction is controlled with pressure regulators. For a discussion of high-pressure transmission of gaseous hydrogen, see Volume II, Chapter 1 of this series.

Before the gas leaves a city gate station, a small, controlled amount of substance with a strong, penetrating odor is injected into it with an odorizer. This odorant warns customers of the presence of the unburned gas before it accumulates in hazardous quantities. The piping that picks up the gas at the city gate station and carries it to the remainder of the distribution system is a supply main. After the supply main, the piping in a distribution system can be categorized as:[3]

1. Trunk mains to supply gas from a major source, such as a supply main, to feeder mains
2. Feeder mains to supply gas from trunk mains to distributor mains, as well as to the services connected to them
3. Distribution to mains to supply gas primarily to services
4. Services to deliver gas from a main to the customers' gas meters

Many distribution systems comprise several superimposed networks of piping oper-ated at different pressure levels. Gas from the "high-pressure" supply mains can be fed through pressure regulators into high-, medium-, or low-pressure distribution net-

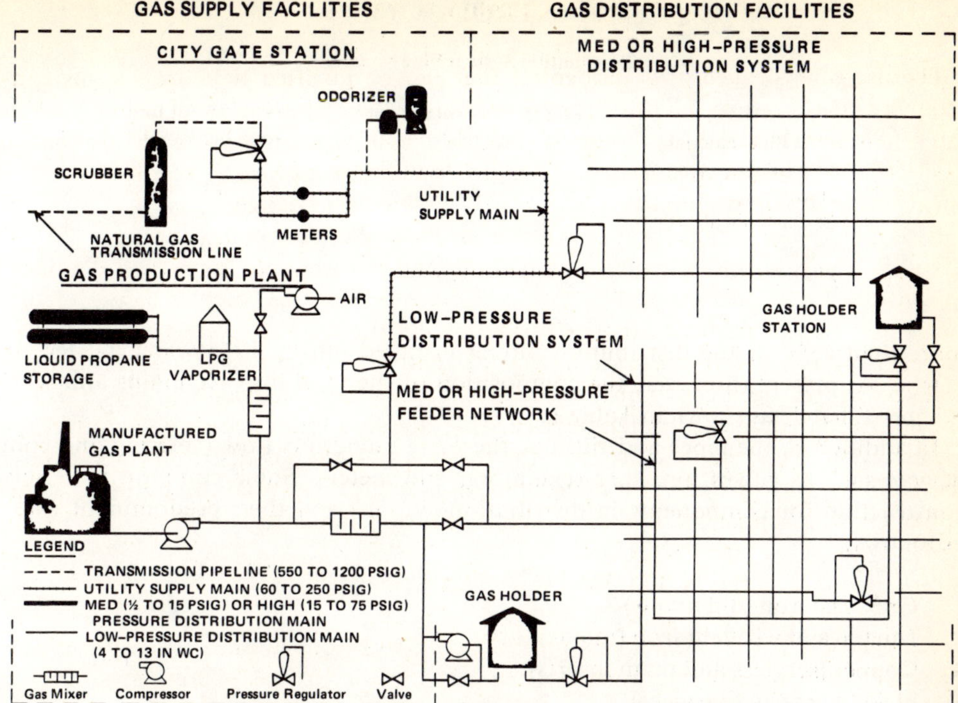

FIGURE 1.   A gas distribution system with representative supply sources. (From IGT Home Study Course — Gas Distribution, Institute of Gas Technology, Chicago, 1963. With permission.)

works. High-pressure distribution networks, operated at up to 75 psig, are of two types:

1. Networks that serve customers directly (with large quantities of gas)
2. Networks used primarily to feed gas into medium pressure (0.5 to 15 psig) or low-pressure (4 to 12 in. H_2O gauge) distribution networks

The fuel lines on the premises of residential customers are normally operated at low pressure measured in inches of water (gage); the pressure required for proper appliance performance usually is controlled by a regulator on the appliance. The purpose of the service regulator in each residence (fed by service lines from the distribution mains) is to reduce the gas pressure to that of the fuel line.

The basic distribution system in most older, large cities is a low-pressure cast-iron pipe system. Most of the piping in these systems was laid during the manufactured-gas era when gas (containing up to 50% hydrogen) was produced at low pressure and when compression was a relatively expensive operation. With higher pressure natural gas now available, major portions of new distribution systems are medium- or high-pressure networks of steel. Consequently, most older low-pressure systems are now overlaid with medium- and high-pressure piping grids.

There is a tremendous diversity of pipes and fittings that has been used, but one trend predominates in newly installed gas distribution systems — the increased use of plastic pipes and fittings for mains of 2 in. and smaller diameters (medium- and low-pressure networks) and for services. According to a recent survey, about 87% of the gas utility companies now install plastic pipes for new and replacement mains, and about 92% use plastic for new and replacement services.[4] On a length or footage basis,

TABLE 1

Distribution System Pipe Materials

Pipe material	Mains, % of total (footage basis)	Services, % of total (number basis)
Steel	80	83
Cast and ductile iron	12	2
Copper	1	6
Plastic	7	9

more than 55% of the distribution piping installed during 1975 was expected to be plastic. At present, however, the composition of the total in-place mains and services is approximately as shown in Table 1.

In addition to the pipes and fittings, there are numerous flow metering and control devices such as valves, pressure regulators, and meters. Some common materials of construction for components in distribution systems and their predominant uses are as follows:

1.   Gray cast iron (old mains)
2.   Ductile and wrought iron (mains)
3.   Copper (services and main inserts)
4.   Steel (mains and services)
5.   Brass (valves)
6.   Polyethylene (services and mains)
7.   Fiberglass-reinforced epoxy (services and mains)
8.   Natural and synthetic rubber, elastomers (mechanical joints seals, meter diaphragms)
9.   Lead and jute (sealer, bell, and spigot joints)
10.   Cast aluminum (meter housings and regulator parts)
11.   Cadmium-plated steel (internal regulator parts)
12.   Cork and gasket materials (meter and instrument gaskets)
13.   Miscellaneous plastics (ABS, CAB, PB, nylon, etc., services, and internal instrument parts)

The diversity of equipment, operating conditions, and materials of construction shows that compatibility of the distribution system with hydrogen requires verification by demonstration. Any statement that hydrogen or hydrogen-rich gases can be adequately and safely delivered to the customer by using the in-place natural gas distribution system is a presumption until laboratory and field tests prove it to be fact.

3.3. COMPATIBILITY AND PROBLEMS IN HYDROGEN DISTRIBUTION

3.3.1. Volumetric Flow

In general, the operating conditions for a distribution line are determined by "delivery" of the quantity tof hydrogen equivalent (in high heating value) to that of (existing) natural gas. Thus, the flow of hydrogen in proportion to that of natural gas must be increased by a factor of about 2.9 to 3.4, depending upon the natural gas heating value. As a result, some volumetric flow meters used for measuring gas delivery could be undersized for hydrogen usage.

The flow in natural gas distribution systems is usually partially turbulent, and it can be laminar or turbulent depending on location and momentary flow conditions. The (dimensionless) Reynolds number, which mathematically characterizes the flow, depends directly upon pipe diameter, gas velocity, and gas density and inversely upon gas viscosity. For the three flow conditions, the amount of hydrogen and the amount of energy delivered are different for the same pipe with the same pressure drop.

The Reynolds number can be expressed as

$$Re = \frac{Dv\rho}{\mu} \tag{1}$$

where D = pipe diameter, v = average gas velocity, ϱ = gas density, and μ = gas viscosity.

Typical fluid properties for hydrogen and methane (similar to natural gas) are as follows:

	H_2	CH_4
μ (cP)	0.0085	0.011
ϱ (lbm/ft^3)	0.0053	0.042

3.3.1.1. Laminar Flow

For laminar flow (to which $Re < 2000$ usually corresponds), the Poiseuille equation can be applied:

$$\Delta P = K \left(\frac{\mu L v}{D^2}\right) \tag{2}$$

where Δp = gas pressure drop for pipe length L, L = pipe length, and K = constant (depends on units). Applying Equation 2 to both hydrogen and methane, where Δp, D, and L are constant, gives Equation 3:

$$v_{H_2} = 1.29 \, v_{CH_4} \tag{3}$$

Hence, in laminar flow for the same pressure drop in the same pipe, hydrogen travels only about 30% faster than methane (or natural gas).

In terms of the Reynolds numbers for these conditions,

$$\frac{Re_{H_2}}{Re_{CH_4}} = 0.163 \, \frac{v_{H_2}}{v_{CH_4}} \tag{4}$$

To derive Equation 3, we used the standard relationship between friction factor, f, and Reynolds number,

$$f = \frac{64}{Re} \tag{5}$$

Alternatively, an appropriate (laminar) transmission factor could be used with a general pipeline flow formula.[5] Transmission factors, those to be used for partially turbulent and turbulent flow, were developed for natural gas. Their application to hydrogen is not strictly valid, but the alternative — ignoring them in these comparisons — is not desirable. For laminar flow, volumetric flow rates will be about proportional to the average velocities of hydrogen and natural gas per Equation 3.

In terms of energy delivery, the high heating value of hydrogen is 325 Btu/scf, the high heating value of methane is 1012 Btu/scf, and the high heating value of a typical natural gas is 1060 Btu/scf. Hence, for laminar flow in a given distribution pipe under the same pressure drop,

$$H_2 \text{ energy delivery} = 0.41 \ CH_4 \text{ energy delivery}$$

$$H_2 \text{ energy delivery} \approx 0.40 \text{ natural gas energy delivery}$$

3.3.1.2. Partially Turbulent Flow

For a given pipe, partially turbulent flow exists when the fluid velocity ranges between that of laminar flow and fully developed turbulence. Usually, the Reynolds number for this condition is above 2000 to about 10^5; the limits are very dependent on pipe roughness and irregularities in pipe configuration. It is partially turbulent because of the laminar sublayer near the pipe wall.

The Darcy general equation for fluid flow can be used for this flow condition:

$$\Delta P = k \ \left(\frac{\rho f L v^2}{D}\right) \tag{6}$$

where Δp = gas pressure drop for pipe length L, L = pipe length, D = pipe diameter, ϱ = gas density, v = average gas velocity, f = friction factor, and k = constant.

In order to apply Equation 7, the friction factors for hydrogen and methane must be known, but the friction factor also is a function of gas velocity. Assuming that the friction factor is a weak function of Reynolds number in the partially turbulent flow region,

$$f_{H_2} = 1.2 \ f_{CH_4} \tag{7}$$

For the same pipe where the Δp, D, and L are constant, the relationship of Equation 9 results for partially turbulent flow:

$$v_{H_2} = 2.6 \ v_{CH_4} \tag{8}$$

In this case, the average velocity of hydrogen is about three times that of methane.

The relationship between the Reynolds numbers for partially turbulent flow is

$$Re_{H_2} = 0.42 \ Re_{CH_4} \tag{9}$$

In terms of volumetric flow rates, the amount of gas delivered is different, primarily because of the difference in specific gravity (density) and Reynolds number. For gas distribution systems, the difference in compressibility factors for hydrogen and methane is negligible. The volumetric flow rate, q, can be calculated by Equation 5:

$$q = k \ (\text{compressibility factor} \times \text{specific gravity})^{-0.5} \ (\text{transmission factor}) \tag{10}$$

where k = constant, compressibility factor $\approx$ 1.0, specific gravity H_2 = 0.07 (air = 1.0), specific gravity CH_4 = 0.55, and (partially turbulent) transmission factor = 4.17 $(Re)^{0.10}$

By using Equation 10,

$$q_{H_2} = 2.63 \ q_{CH_4} \tag{11}$$

By using the heating values, the relative energy deliveries can be calculated from Equation 11:

$$H_2 \text{ energy delivery} = 0.84 \; CH_4 \text{ energy delivery}$$

$$H_2 \text{ energy delivery} \approx 0.80 \text{ natural gas delivery}$$

This treatment for partially turbulent flow assumes that there is a gradual transition, not an abrupt change, in the shape of the Reynolds number vs. friction factor curve in which the flow condition passes from laminar into partially turbulent flow. If this is not true for the case of methane (natural gas) vs. hydrogen, Equation 8 does not hold, but the results are not severe. In fact, an approximate relationship such as

$$f_{H_2} = 0.5 \; f_{CH_4} \tag{12}$$

is not impossible, and it would result in the following estimates:

$$v_{H_2} = 2 \; v_{CH_4} \tag{13}$$

$$Re_{H_2} = 0.3 \; Re_{CH_4} \tag{14}$$

$$q_{H_2} = 2.5 \; q_{CH_4} \tag{15}$$

$$H_2 \text{ energy delivery} = 0.8 \; CH_4 \text{ energy delivery}$$

3.3.1.3. Turbulent Flow

For fully developed turbulent flow, the Darcy equation again applies, and a different transmission factor is more appropriate:[6]

$$\text{(turbulent) transmission factor} = 4 \log (3.7 \; D/k)$$

For this flow condition in the same pipe under the same pressure drop, the following results:

$$f_{H_2} = 1.0 \; f_{CH_4} \quad \text{(not a function of Re)} \tag{16}$$

$$v_{H_2} = 2.95 \; v_{CH_4} \tag{17}$$

$$Re_{H_2} = 0.48 \; Re_{CH_4} \tag{18}$$

$$q_{H_2} = 2.82 \; q_{CH_4} \tag{19}$$

$$H_2 \text{ energy delivery} = 0.90 \; CH_4 \text{ energy delivery}$$

$$H_2 \text{ energy delivery} \approx 0.86 \text{ natural gas energy delivery}$$

These preliminary calculations show that the operating pressures (and pressure drops) will most likely be increased in distribution systems to deliver the same energy with hydrogen as is now delivered with natural gas. Primarily to increase gas density, but also to increase average velocity, operating pressures for hydrogen would be higher than with natural gas by about 10 to 25%.

3.3.2. Odorants and Illuminants

As far as is known, the sulfur compounds now used to odorize natural gas are compatible with hydrogen and would be satisfactory. Odorants have to be added on a volume basis. Thus, because about three times the volume of hydrogen is to be used, three times the quantity of odorant will be required, although its actual percentage level will be the same. The odorant would be burned at the appliance.

When an odorant is being introduced into the gas, the simultaneous addition of an illuminant to make the hydrogen flame visible would also be worthwhile. This would facilitate the use of hydrogen in open-flame appliances and assist in the adjustment of burners and pilots. No suitable illuminant has yet been identified, but two classes of materials are appropriate. One is a small amount of an aromatic organic material that would burn with a yellow flame; the other is a trace of a volatile organometallic sodium compound that would give the flame the characteristic sodium-yellow color. Experiments are required to determine whether either of these approaches would be useful and to determine the optimum quantities to be added.

The addition of any foreign material, such as sulfur-bearing odorants, could cause considerable problems with catalyst poisoning if hydrogen is used in catalytic burners or processes. It is technically possible to remove the odorant immediately before the catalysis, but in some applications this would be inconvenient and would add cost. An alternative, worthy of consideration for future experimental work, is to develop an odorizing material that does not interfere with catalysts.

3.3.3. Leakage

The rate of loss of hydrogen through orifices, cracks, corrosion pinholes, or leaky seals is about 2½ to 3 times greater on a volume basis than that for methane or natural gas. However, the rate of energy loss is about the same as for natural gas. If the leak is very small, e.g., a hairline crack, so that the primary mechanism of transport is equivalent to molecular diffusion, the relative leak rate for different gases is inversely proportional to the square root of the molecular weights of the gases:

$$\text{Leakage ratio} \ \frac{H_2}{CH_4} = \sqrt{\frac{M_{CH_4}}{M_{H_2}}} \tag{20}$$

where M_{CH_4} = molecular weight methane (16), and M_{H_2} = molecular weight hydrogen (2).
From Equation 20,

$$\text{Leakage ratio} \ \frac{H_2}{CH_4} = 2.83 \ (\text{diffusion}) \tag{20a}$$

The rate of loss of hydrogen through a hole or orifice to the atmosphere, compared with that for natural gas, assuming the same pressure drop conditions, can be approximated by using Equation 21, which can be derived from the Bernoulli theorem.

$$q = YCA \ [(2 g \ (144) \ \Delta p)/\rho] \ ^{0.5} \tag{21}$$

where q = flow rate, ft³/sec, Y = hydrogen expansion factor, 0.45,[6] Y = methane expansion factor, 0.50,[6] A = cross-sectional orifice area, $A(H_2) = A(CH_4)$, C = discharge coefficient, $C(H_2) = C(CH_4) \pm 2\%$,[7] ρ = hydrogen density, 0.0053 lbm/ft³, and ρ = methane density, 0.047 lbm/ft³.
Substituting into Equation 21 yields

$$q_{H_2} = 2.54 \ q_{CH_4} \tag{22}$$

or,

$$\text{Leakage ratio} \ \frac{H_2}{CH_4} = 2.54 \ (\text{laminar flow through orifice}) \tag{23}$$

TABLE 2

Permeability of Plastic Piping Compounds

Material	Gas permeability, 10^{-3} CF mil/ft^2-day-atm		Relative permeability, hydrogen/methane
	Methane	Hydrogen	
Acrylonitrite-butadiene styrene, ABS-1,2	0.59	51.71	87.6
Cellulose acetate-butyrate, CAB-MH	11.8	157.0	13.3
Polypropylene	1.6	20.2	12.6
Polyethylene PE III-3	2.3	15.7	6.8
Polyvinyl chloride, PVC-II-I	0.2	13.7	68.5

Note that, on an energy basis, for a leak through a hole in a distribution pipe, about 20% *more* methane energy would be lost than hydrogen energy. For natural gas, exact and uniformly applicable values of coefficients Y and C are not available, but for different natural gases, these values would not be significantly different than the values for methane.

For normal purposes, cast iron and steel pipes are almost impermeable to both hydrogen and natural gas at ambient temperatures. In fact, even though the permeability rate of hydrogen is higher than that of natural gas, it is still so small that it is insignificant. Similar reasoning should apply to sealing materials, but, to be sure, experimental confirmatory tests should be carried out on typical seals.

The permeability of plastic pipes to hydrogen, however, is high enough to warrant further consideration. Some permeability measurements of typical plastic piping compounds have been made. In these studies, the ratio of permeability of hydrogen to that of natural gas varies widely — between about 7:1 for polyethylene and 88:1 for ABS. Table 2 shows some relative permeability data measured by Battelle Memorial Institute.[8]

In comparison with typical leakage rates for natural gas distribution systems, the hydrogen lost by permeation (through plastic) appears insignificant; in volumetric terms, it would be about 2 orders of magnitude less than the natural gas loss now tolerated due to (accepted) leaks.

In meters and regulators, containment of hydrogen might also be a problem. Polymeric materials and elastomers are used for seals (gaskets) and for diaphragms in regulators (for example). Regulators now in use are constructed of parts selected for natural gas service. Other materials probably would be safer and could exhibit better performance for hydrogen service. A regulator diaphragm that allows hydrogen to permeate through the edge to the atmosphere is not desirable.

3.3.4. Peculiar Temperature Effects

Unlike methane or natural gas, hydrogen has a maximum Joule-Thompson inversion temperature of $-109°$F, far below ambient temperatures; therefore, at ambient temperatures, hydrogen undergoes a temperature increase upon throttled expansion. The extent of the observed temperature rise depends upon the pressure drop, the initial gas temperature, and the rate of heat exchange between the hydrogen and the environment. This Joule-Thompson effect could be significant in plastic components. Thermoplastic pipe for natural gas service, for instance, is not required to be pressure tested (at the greater of 50 psig or 150% of operating pressure) at temperatures above 100°F.[9]

3.3.5. Line Purging and Maintenance

An area of serious concern is the purging requirements for pipes, both when bringing new mains into service and when repairing existing mains. Use of an inert gas to sweep out the air, or to sweep out the gas already in a used main, may be necessary because the wide flammability limits of hydrogen (4 to 75%), compared with those of natural gas (5 to 15%), would make a hazardous condition in the main more common. With natural gas, welding operations can and are carried out when the gas-air mixture in the line is well above 15% gas. However, with hydrogen, the hydrogen/air ratio would have to be safely above 75%. Use of some form of oxygen-level indicator may be necessary to ensure safe conditions. Purging procedures for hydrogen may be advantageous and would have to be established by experiment.

3.3.6. Metallurgical Effects

Although hydrogen degradation of metals has been recognized in the failure of certain structures, the exact mechanism by which this occurs is not clearly understood. (For a more complete discussion of this subject, see Volume II, Chapter 4 of this series.) Three categories to explain the problem have been proposed:

1.   Hydrogen-reaction embrittlement
2.   Internal-hydrogen embrittlement
3.   Hydrogen-environment embrittlement

Hydrogen-reaction embrittlement is the result of chemical reaction between hydrogen and a metal or some alloy. Some examples are the formation of irreversible hydrides of titanium, zircronium, and tantalum. The decarburization of steels and the formation of high-pressure water bubbles and methane in metal voids are other examples of hydrogen-reaction embrittlement.

Both internal and environmental hydrogen embrittlement are due to hydrogen atoms dissolving in the metal. Internal hydrogen embrittlement can be produced by any hydrogen-containing chemical solution. This problem is often encountered in metal- or petrochemical-processing facilities.

Hydrogen-environment embrittlement is the degradation of mechanical properties due to the adsorption of hydrogen at the surface of a metal. This type of embrittlement is manifested when metal crack formation initiates in the presence of hydrogen. Rate of strain is an important parameter in the degree of embrittlement exhibited. Low strain rates promote maximum embrittlement, allowing hydrogen transport near the cracks. Metals conditioned to high strength levels are often more susceptible to embrittlement than their lower strength counterparts.

These modes of hydrogen attack have been observed to occur under pressure and temperature conditions far more extreme than those of a gas distribution system. These metallurgical effects would not be expected to occur in distribution equipment used for hydrogen service because of the low operating pressures and ambient temperatures.

3.4. RESIDENTIAL ENERGY USE PATTERNS

Assessed in this section is the feasibility of using hydrogen to supply a substantial portion of the energy requirements for a single-family dwelling unit. This assessment of residential hydrogen utilization is based on the schematic diagram presented in Figure 2. It is made from the point at which energy enters the residential unit and will not be concerned with the factors affecting energy production, transmission, and distribution. As seen in Figure 2, we are concerned with both the present consumer of

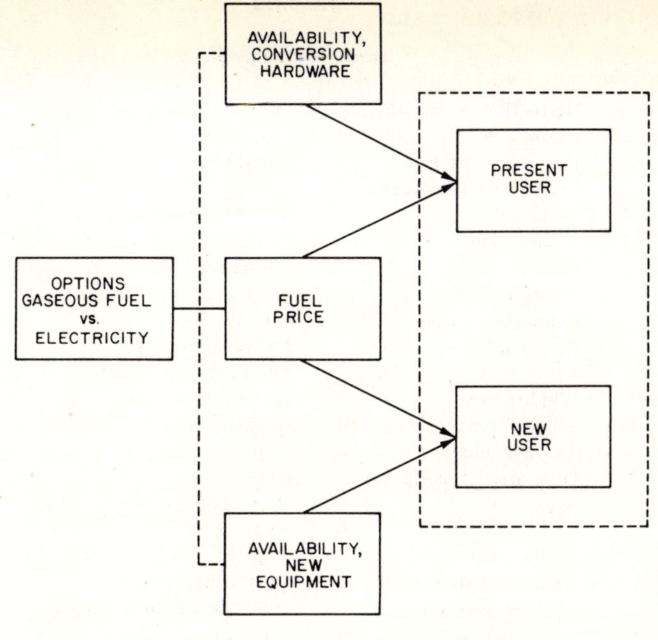

FIGURE 2.   Gaseous fuel vs. electricity.

energy and the new customer. The present energy user already owns appliances and is faced with the probability of converting these appliances from present energy forms to new ones as the conventional fossil-fuel forms of energy are depleted. The decision to convert will be based on such factors as the age of a particular appliance, the cost of conversion, and the cost of alternative energy supplies.

The new energy user does not currently own any appliances but instead has to decide which to buy. The purchase decision is likely to be based on the cost and type of energy available to him and on the initial cost of the equipment. One exception to this is the availability of solar heating. It is assumed that such systems will be available within the near-term time frame; however, the cost of such a system is expected to be high, while the related energy or fuel cost will be zero (except for possible standby equipment). Should the new customer choose some form of ready-built housing, the decision as to a specific configuration of appliances will already have been made for him.

A prerequisite to a discussion of energy utilization in the residential market is knowledge of the quantities of various energy forms consumed for such purposes as space heating, cooling, cooking, water heating, clothes washing and drying, and lighting. For purposes of discussion and later comparison, a specific dwelling unit type was chosen for use in quantifying residential energy consumption.[18] The consumption is not to be considered as representative of patterns across the U.S. but merely serves as an example. This configuration, for a typical, single-family dwelling unit in the northeastern U.S., was developed by the Bureau of Census from data collected on housing construction techniques within that region. Table 3 is the resulting list of housing parameters. It can be seen that the unit outlined is dependent upon both natural gas and electricity as energy sources. Annual energy consumption within the above residence is shown, by application, in Table 4. The energy consumed is predicated on installation of the most modern appliances found in a newly constructed home.

Data on residential appliance gas consumption are available from a number of

TABLE 3

Housing Parameters

House floor area	1500 ft² (finished)
House style	Two story
House construction	Wood frame
Exterior-wall cnstruction	
Surface	Wood shiplap
Sheathing	½-in. insulation board
Insulation	R-7 batting
Inside	½-in. dry wall
Ceiling insulation	5 in. blown in
Basement type	Full (unfinished)
Attic	Ventilated, unheated
Window area	12% of floor area
Window type	Al casement
Storm windows	None
Door area (three doors)	60 ft²
Door type	Wood panel with 0.5 ft² of glass pane
Storm-door area	40 ft²
Patio door area	40 ft² (single pane)
Window covering	70% draped; 20% shaded; 10% open
	No awnings
External landscaping	No shading effect
Direction house faces	North
External colors	White roof and walls
Roof construction	Asphalt shingle
Heating system	Forced hot air, natural gas
Cooling system	Central, electric
Garage (enclosed)	Attached, slab, unheated
Residents	Two adults, two children
Location	Northeastern region of U.S.

Note: In addition to the above housing characteristics, external weather conditions must be defined.

sources. In this discussion, we have chosen to use some information made available by a series of tests performed by Northern Natural Gas Co. of Omaha, Nebraska.[25] These data were selected because they are presented as a function of water temperatures for the ease of water-heating consumption. This method of data presentation makes the data useful and readily convertible to other geographic regions. The test data are actual measurements made in residential customers' homes through the use of specially installed meters that were monitored over a period of time.

It can be seen in Table 4 that space and water heating are the major energy-consuming applications within the given residential situation. It should also be noted that the quantity of electricity consumed within a residence can vary by a factor of 2 or more for a specified floor area. This variance in consumption can be primarily attributed to assorted living patterns, human factors, and the number and type of appliances within a particular situation.

3.4.1. Space Heating

Space heating is the single largest energy-consuming application within the example dwelling unit. As shown in Table 4, it accounts for 64.8% of the total energy consumed.

TABLE 4

Annual Energy Consumption in Residential Unit, Northeast Region

	Energy consumed (10⁶ Btu)	% of total consumption
Natural gas load		
Space heating	132.0	64.8
Water heating	27.0	13.3
Cooking	6.4	3.1
Clothes drying	3.4	1.7
Subtotal	168.8	82.9
Electric load		
Central air conditioning	12.5	6.1
Lighting	6.8	3.3
Refrigerator-freezer	6.2	3.1
Clothes washer	0.3	0.2
Color TV	1.7	0.8
Furnace fan	1.3	0.6
Dishwasher	1.2	0.6
Iron	0.5	0.2
Coffee maker	0.4	0.2
Miscellaneous	4.1	2.0
Subtotal	35.0	17.1
Total	203.8	100.0

The estimated annual gas consumption for space heating in single-family dwellings, by size, is presented in Figure 3. Data are presented for house sizes ranging from 800 to 2000 ft² of floor area and for annual degree days* ranging from 5000 to 10,000. The estimated annual gas composition can be seen to vary from 100,000 to 280,000 scf. The estimated peak-day requirements by house size at an average daily outdoor temperature of 0°F are presented in Table 5. These data are useful in illustrating the changes in consumption levels for dwelling units of various sizes.

The standing pilot flames associated with present residential space-heating equipment consume about 20 to 35 scf of natural gas per day. At the average rate of 27 scf per day, the heating unit will have an annual pilot-light load of approximately 10,000 scf if the pilot burns throughout the entire year. The pilot-light load is included in the quantity of annual energy consumed listed above.

3.4.2. Water Heating

The annual quantity of energy required for water heating is more dependent upon the number of occupants in a home than on the number of degree days. The estimated water-heating load is presented in Table 4 and is equivalent to about 13.3% of the

---

* Degree-day heating: a measure of the coldness of the weather experienced based on the extent to which the daily mean temperature falls below a reference temperature, usually 65°F. For example, on a day when the mean outdoor dry-bulb temperature is 35°F, there would be 30 degree days experienced. A daily mean temperature usually represents the sum of the high and low readings for the day divided by two.

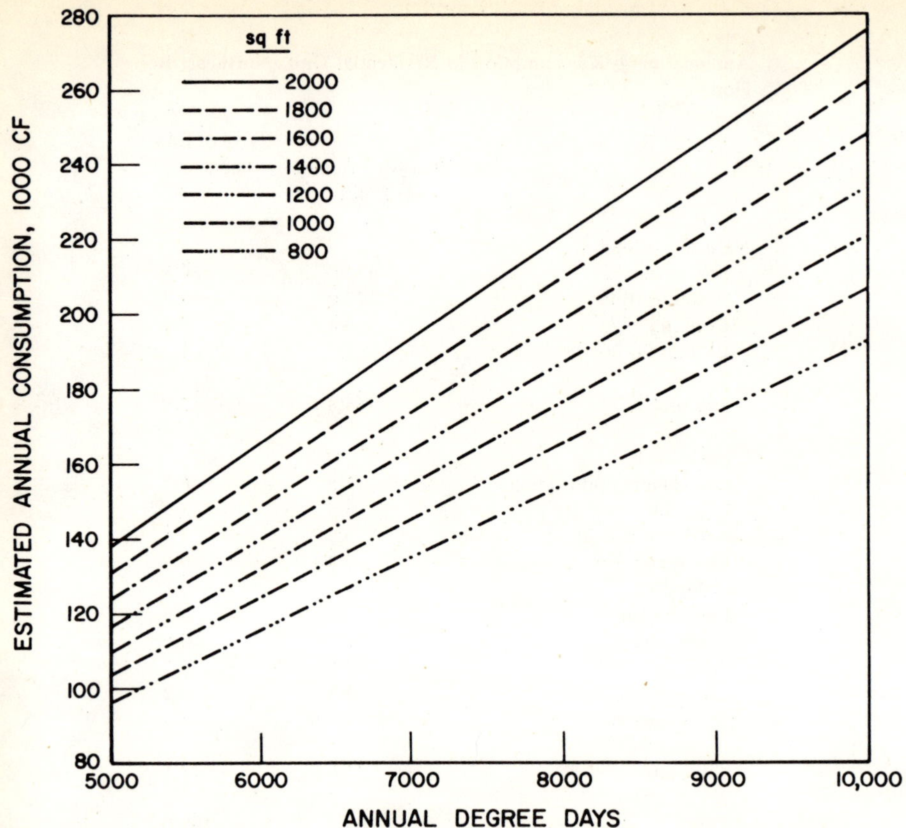

FIGURE 3.   Estimated annual consumption by dwelling unit size.

<table>
<tr><td colspan="2" align="center">TABLE 5</td><td colspan="2" align="center">TABLE 6</td></tr>
<tr><td colspan="2">Estimated Peak-Day Natural Gas Requirements, by House Size at an Average Daily Temperature of 0° F</td><td colspan="2">Annual Water-Heating Load, by Family Size</td></tr>
<tr><td>House size (ft²)</td><td>Peak day (SCF)</td><td>Occupants (no.)</td><td>Estimated load (SCF)</td></tr>
<tr><td></td><td></td><td>1</td><td>32,820</td></tr>
<tr><td></td><td></td><td>2</td><td>35,660</td></tr>
<tr><td>800</td><td>1302</td><td>3</td><td>38,500</td></tr>
<tr><td>1000</td><td>1398</td><td>4</td><td>41,336</td></tr>
<tr><td>1200</td><td>1494</td><td>5</td><td>44,176</td></tr>
<tr><td>1400</td><td>1590</td><td>6</td><td>47,016</td></tr>
<tr><td>1600</td><td>1685</td><td>7</td><td>49,855</td></tr>
<tr><td>1800</td><td>1781</td><td>8</td><td>59,691</td></tr>
<tr><td>2000</td><td>1877</td><td></td><td></td></tr>
</table>

total energy consumption. It can be seen in Table 6 that annual energy consumption varies directly with the number of family members. For the family of four presented in the example dwelling unit, the energy consumed on an annual basis is about 41,000 scf.

There is a seasonal variation in the amount of energy required that is related to the temperature of the incoming water. Because the city water supply is warmer in the summer than in the winter, less heat is required to bring it to the desired thermostatically controlled temperature. This variation is shown in Figure 4. It can be seen that

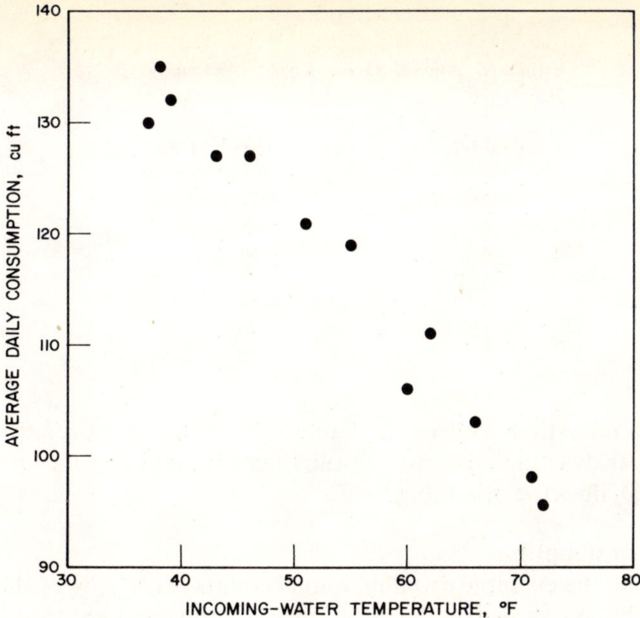

INCOMING-WATER TEMPERATURE, °F

FIGURE 4.   Comparison of average monthly water temperature and
average daily water-heater consumption, by month.

water-heating loads peak in the winter season at a value approximately 30% above the summer level.

The pilot-light load for water heating with natural gas is estimated at about 17 scf per day, or 6200 scf per year. (These loads are included in the estimates presented in Table 4.) The range is 24 scf per day, with the majority of the units consuming between 15 and 20 scf per day.

3.4.3. Cooking

The natural gas requirement for cooking in our example dwelling unit is about 6.4 million Btu/year. This is equivalent to 3.1% of the total energy consumed, as shown in Table 4. The pilot-light consumption is a major factor in cooking ranges and accounts for about 60% of the total amount, or 2.4 million Btu. The magnitude of the annual gas load for use in cooking is difficult to relate to family size because of existing variations in individual cooking habits.

The peak-day cooking load occurs, as might be expected, on either Christmas or Thanksgiving Day. Some heavy cooking loads also occur during July and August and can be attributed to the canning season.

3.4.4. Clothes Drying

The estimated annual natural gas requirement for clothes drying, as in the case of water heating, is dependent upon the number of occupants in the dwelling. As shown in Table 4, the amount of energy consumed in clothes drying is about 3.4 million Btu, or 1.7% of the total energy consumption. A tabulation of estimated annual gas loads for clothes drying, based on the number of occupants, is shown in Table 7.

The estimated loads shown in Table 7 do not include pilot-light consumption because gas pilot flames are used only in a small number of gas dryers currently in use. Generally, electric ignition systems are installed on the more expensive gas dryers. For dryers without electric ignition of the dryer flame, an annual load of approximately 3000 scf

TABLE 7

Estimated Annual Dryer Load, by Family
Size

Occupants (no.)	Estimated load (SCF)
2	2321
3	3391
4	4464
5	5533
6	6606
7	7680
8	8740

should be added to the estimates listed in Table 7. It should be noted that, in the case of two occupants per dwelling unit, the pilot light consumes more natural gas over the period of a year than does the operating load.

3.4.5. Electricity Consumption

The electric load in the example dwelling unit accounts for 17.1% of the total annual energy consumption. As shown in Table 4, this is equivalent to 35 million Btu, or 10,255 kWh. As in the case of cooking, the consumption of electricity is dependent upon many variables that cannot be measured as accurately as degree days, nor can it be based simply on the number of occupants.

At present, the amount of electricity consumed within a residence can vary by a factor of 2 for any given floor area. The increase in the use of electricity within a residential unit can be illustrated by the history of New York state usage over 10 years (from 1960 to 1970) (see Figure 5). The annual electrical use per customer has almost doubled. This trend has been created by the addition of appliances that are principally conveniences. The quantity of electricity consumed by various appliances on an annual basis is shown in Table 8.

3.5. THE USE OF HYDROGEN IN DOMESTIC APPLIANCES

If hydrogen is to replace natural gas in residential appliances, the behavior of hydrogen in burners designed to combust natural gas must be explored. The following discussion is intended to show why hydrogen cannot be directly substituted into residential appliances and what burner modifications would be necessary before hydrogen could be used as a fuel. It should be kept in mind that similar equipment modifications were necessary when natural gas was substituted for manufactured gas.

Hydrogen has a number of combustion properties that can be beneficially exploited with burners that operate through the action of a catalyst (i.e., catalytic burners) rather than through the action of a flame (i.e., atmospheric burners). The experimental work performed to date on such burners will also be described.

3.5.1. Conversion of Existing Appliances
3.5.1.1. Atmospheric Burners

The burners built into domestic gas appliances are designed to burn fuel from a low-pressure (from 3 to 16 in. water column) gas source, so they are known as "atmospheric burners". These burners operate on the same principles as a Bunsen burner.

A Bunsen burner consists of a straight, smooth metal tube with a gas-metering orifice at the lower end. Ambient air enters the tube through adjustable openings around

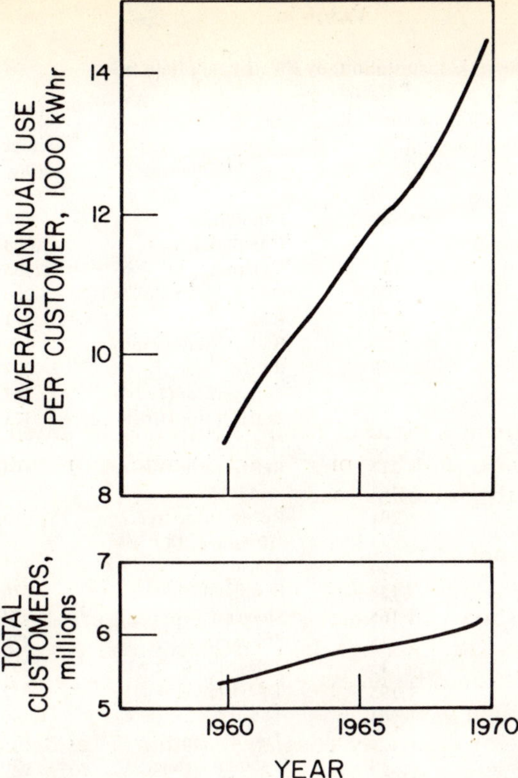

FIGURE 5.   Historical electricity growth, New York state.

the gas orifice and is transported (entrained) by the high-velocity gas stream (jet). The air-gas mixture is ignited as it emerges from the upper end of the tube (the burner port). The air supplied through the burner openings near the metering orifice, before combustion, is *primary air*. Ambient air mixed after ignition is *secondary air*. Figure 6 shows the basic construction and nomenclature of an atmospheric gas burner.

During operation, gas issues at a high velocity from the gas orifice (which also meters the gas flow into the burner), creating a vacuum or lowered pressure, which allows the primary air to enter and mix with the gas. While continuing to mix, the air-gas mixture moves down the tube and into the burner head. The mixture passes through the ports and is ignited. The amount of primary air is described as a percent of the theoretical air required for complete combustion. For example, 9.56 ft^3 of air is required to burn 1 ft^3 of methane (CH_4):

$$CH_4 + 2O_2 + 7.56N_2 \rightarrow CO_2 + 7.56N_2 + \text{heat} \qquad (24)$$

Therefore, a 65% primary air mixture would have 6.214 ft^3 of air mixed with 1 ft^3 of methane. The balance of the air required to complete combustion, i.e., 3.346 ft^3, would then be secondary air. Figure 7 shows flame geometry as a function of percent of primary air and natural gas.

For efficient operation, atmospheric burners must be able to perform under a wide variety of field conditions and must meet the following basic requirements:[14]

TABLE 8

Energy Consumption, by Electric Appliances

Appliance	Average wattage rating	Estimated annual consumption, (kWhr)	Appliance	Average wattage rating	Estimated annual consumption, (kWhr)
Air conditioner[a] (window)	1,566	1,389	Humidifier	117	163
			Iron (hand)	1,008	144
Bed covering[a]	177	147	Iron (mangle)	1,465	165
Broiler	1,436	100	Oil burner or stoker	266	410
Carving knife	92	8	Radio	71	86
Clock	2	17	Radio-phonograph	109	109
Clothes dryer	4,856	993	Range	12,207	1,175
Coffee maker	894	106	Refrigerator (12 ft³)	241	725
Cooker (egg)	510	14	Refrigerator (frost-	321	1,217
Deep-fat fryer	1,448	83	less, 12 ft³)		
Dehumidifier[a]	257	377	Refrigerator-freezer	326	1,137
Dishwasher	1,201	363	(14 ft³)		
Fan (attic)	370	291	Refrigerator-freezer	615	1,829
Fan (circulating)	88	43	(frostless, 14 ft³)		
Fan (furnace)[a]	292	394	Roaster	1,333	205
Fan (rollabout)	171	138	Sewing machine	75	11
Fan (window)[a]	190	165	Shaver	14	18
Floor polisher	305	15	Sun lamp	279	16
Food blender	386	15	Television (b & w)	237	362
Food freezer (15 ft³)	341	1,195	Television (color)	332	502
Food freezer (frost-	440	1,761	Toaster	1,146	39
less, 15 ft³)			Toothbrush	7	5
Food mixer	127	13	Vacuum cleaner	630	46
Food waste disposer	445	30	Vibrator	40	2
Frying pan	1,196	186	Waffle iron	1,116	22
Germicidal lamp	20	145	Washing machine	512	103
Grill (sandwich)	1,161	33	(automatic)		
Hair dryer	381	14	Washing machine	286	76
Heat lamp (infrared)	250	13	(nonautomatic)		
Heater (radiant)	1,322	176	Water heater[b]	2,475	4,219
Heating Pad	65	10	Water pump	460	231
Hot plate	1,257	90			

[a] Estimated for season (6 months) of peak use.
[b] Based on special water heating rate.

From Edison Electric Institute, Marketing Division, New York, 1969.

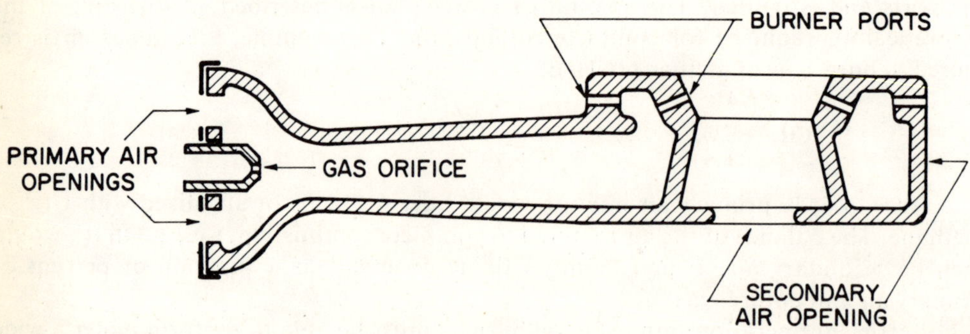

FIGURE 6. Principal parts of a typical atmospheric burner.

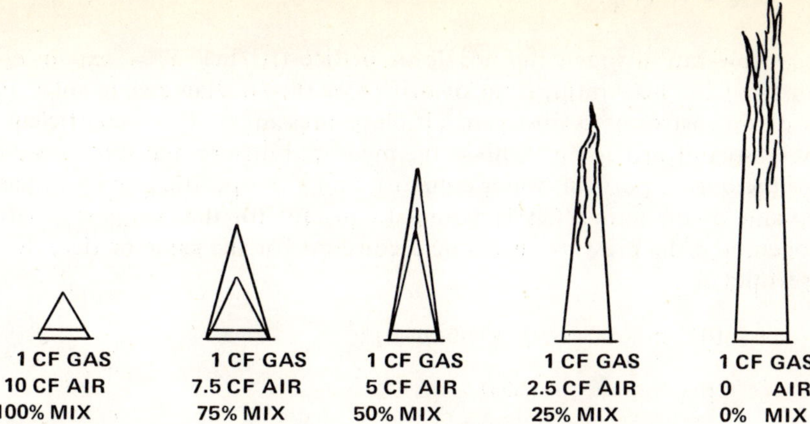

FIGURE 7.   Flame geometry vs. percent aeration. (From *Combusion Technology Manual*, 2nd ed., Industrial Heating Equipment Association, Washington, D.C., 1974. With permission.)

1.   Controllability over a wide range of turn down without danger of flash-back or flame-out
2.   Uniform distribution of heat, including uniform flame height and good flame distribution over the area being heated
3.   Complete combustion — i.e., no formation of carbon (soot) or carbon monoxide (CO)*
4.   No lifting of the flame away from the ports
5.   Ready ignition, i.e., the flame traveling rapidly and without difficulty from port to port over the entire burner
6.   Quiet operation upon ignition, during burning, and upon extinction

Contemporary atmospheric burners are capable of efficient operation over a wide range of input pressures. Range burners, for example, will operate at input pressures ± 50% of the normal pressure of 7 in. water column (0.253 lb/in.²).[11] Atmospheric burners can also be designed to burn any of a variety of fuel gases, some properties of which are shown in Table 9. Tables 10 and 11 give the physical and operating descriptions of the representative atmospheric gas burners used on residential appliances and a calibration-type Bunsen burner shown in Figures 8 and 9, respectively.

The following discussion compares the performance of unmodified burners on natural gas and on hydrogen and shows what modifications should be made during conversion of appliances from natural gas to hydrogen.

3.5.1.1.1. *Fuel Flow Rate*

The gross (high) heating value of a typical natural gas is approximately 1060 Btu/scf, but the gross heating value of hydrogen is only 325 Btu/scf. Thus, for a burner operating on hydrogen to deliver the same amount of heat as one operating on natural gas, per unit time, it will have to pass 3.26 times as much fuel (by volume). Superficially, this would seem to preclude the use of hydrogen in existing appliances without extensive modification. However, the flow of compressible fluids through nozzles (or metering orifices) is governed by the Bernoulli theorem, which can be expressed as

---

*   It should be noted that during start-up, carbon monoxide may be formed if the flame is quenched by a cold target (e.g., water in a water heater).

$$q = YCA \, (h_L/\rho)^{0.5} \tag{25}$$

where q = flow rate through the nozzle or orifice (ft³/hr), Y = expansion factor (function of specific heat ratio, ratio of orifice or throat diameter to inlet diameter, and ratio of downstream to upstream absolute pressures), C = coefficient of discharge, A = area of orifice (in.²), h_L = the measured differential static head or pressure across the burner port (in. water column), and ρ = specific gravity of gas (air = 1). If the value of the term YCA is assumed constant for the two gases, natural gas and hydrogen, and the pressure (h_L) is held constant for the same orifice, the relative flow rates would be

$$q(H_2) = q(\text{nat. gas}) \, [(0.599/0.0696)]^{0.5}$$
$$\tag{26}$$
$$q(H_2) = q(\text{nat. gas}) \, (2.93)$$

Thus the burner, without any changes, will pass 2.93 times as much hydrogen as natural gas. The difference in heat-delivery rate between natural gas and hydrogen is then

$$\frac{(325/1060)}{(1/2.93)} = 0.898 \tag{27}$$

and the hydrogen burner will deliver only about 10% less heat per unit time.

TABLE 9

Gas Mixtures Used with Contemporary Atmospheric-Type Residential Burners

Type of fuel gas	High (gross) heating value, Btu/ft³	Specific gravity (air = 1.0)
Natural	1075	0.65
Manufactured	535	0.38
Mixed	800	0.50
Butane	3175	2.00
Propane	2500	1.53
Butane-air	525	1.16
Butane-air	1400	1.42

TABLE 10

Identification and Description of the Burners in Figure 8

Item no.	Input rating (Btu/hr)	Total no. ports	Total port area (in.²)	Burner construction material	Appliance
1	6,600	48	0.33	Cast iron	Range
2	12,800	64	0.64	Cast iron	Laundry boiler
3	45,000	34	0.63	Cast iron	Water heater
4	9,000	52	0.28	Aluminum	Range
5	9,000	70	0.23	Aluminum	Range
6	12,000	72	0.39	Aluminum	Range
7	5,000	1	0.15	Aluminum	—

TABLE 11

Identification and Description of the Burners in Figure 9

Item no.	Input rating (Btu/hr)	Total no. ports	Total port area (in²)	Burner construction material	Appliance
1	5,000	1	0.15	Aluminum	—
2	9,000	48	0.23	Aluminum	Range
3	30,000	55	1.79	Cold rolled steel	Furnace
4	9,000	69	0.26	Aluminum	Range
5	12,000	72	3.02	Cold rolled steel	Furnace
6	27,500	1	0.91	Cold rolled steel	Water heater
7[a]	40,000	20	0.02	Cold rolled steel	Water heater
8	75,500	42	1.13	Cold rolled steel	Water heater
9	48,000	34	0.63	Cast iron	Water heater
10	6,600	48	0.33	Cast iron	Water heater
11	12,800	64	0.64	Cast iron	Laundry boiler
12	15,000	105	0.76	Cast iron	Broiler

[a] Uses secondary air only.

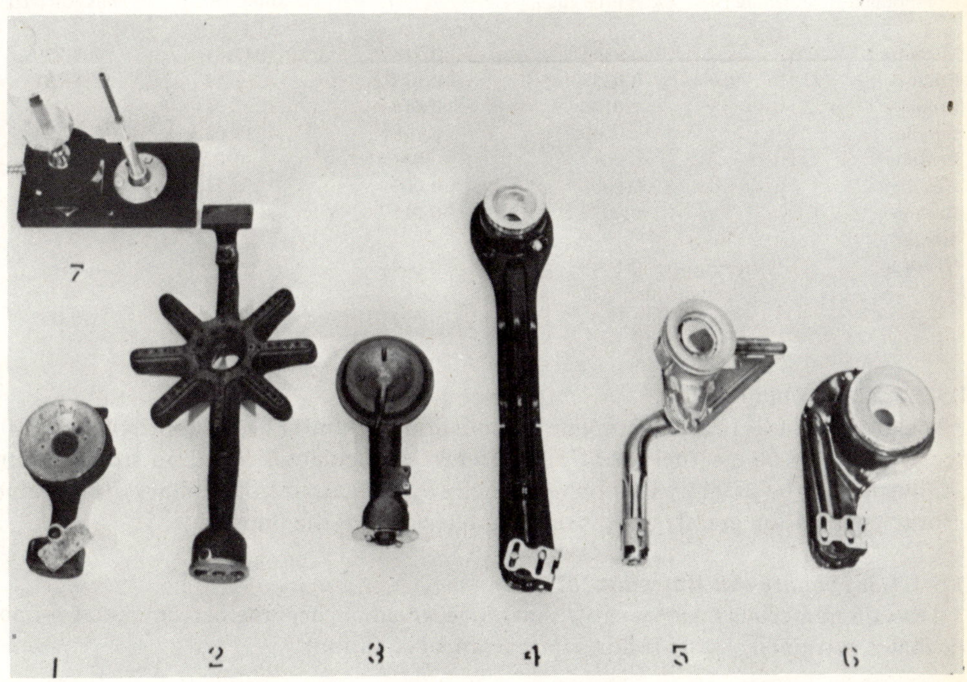

FIGURE 8. Representative atmospheric gas appliance burners (domestic) and a calibration Bunsen burner.

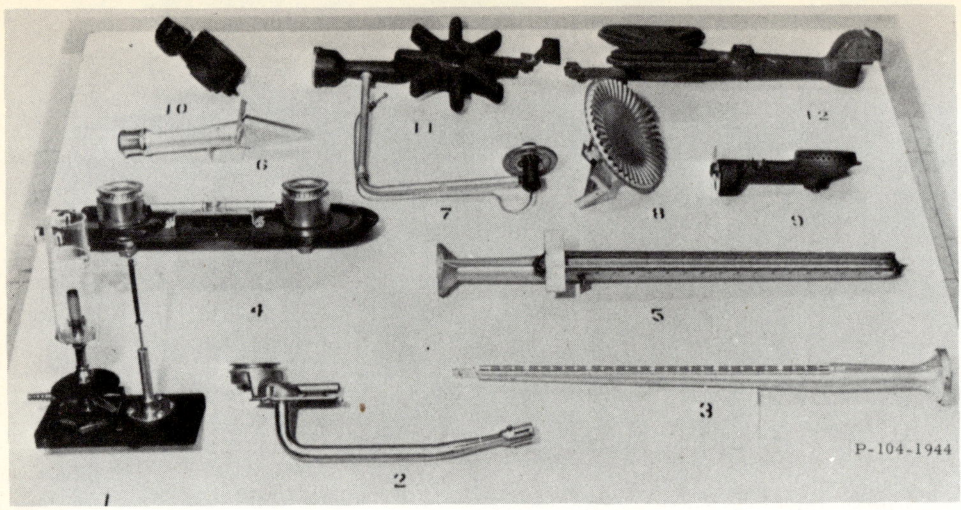

FIGURE 9.   Contemporary gas appliance burners (domestic) with a variety of port constructions and a calibration Bunsen burner.

TABLE 12

Chemical Analysis of a High-Methane Adjustment Gas

Component	Chemical formula	Ft³/Ft³ of gas	Specific gravity × % gas	High (gross) heating value[24] (Btu)	Air required for complete combustion (ft³)
Methane	CH_4	0.9363	0.518	947.84	8.950
Ethane	C_2H_6	0.0358	0.037	63.49	0.597
Propane	C_3H_8	0.0102	0.016	25.74	0.248
Butane	C_4H_{10}	0.0040	0.008	13.06	0.124
Pentane	C_5H_{12}	0.0012	0.003	4.82	0.046
Hexane	C_6H_{14}	0.0008	0.002	3.81	0.040
Carbon dioxide	CO_2	0.0070	0.011	—	—
Nitrogen	N_2	0.0047	0.004	—	—
Total		1.000	0.599	1058.76	10.005

3.5.1.1.2. Air/Fuel Ratio

The air/fuel ratio for the complete combustion of natural gas is approximately 10:1 (see Table 12). The air/fuel ratio for hydrogen is calculated, based on stoichiometric combustion, to be 2.38:1. As shown in Figure 7, the amount of primary air entrained prior to combustion greatly effects the characteristics of the flame.

3.5.1.1.3. Primary-Air Entrainment

Tests on numerous burners show that, for a given burner, the percentage of primary air that is entrained closely follows the empirical equation[14]

$$IP = K[(Pd)^{0.25}/(H)^{0.5}]/R^{0.5} \tag{28}$$

where IP = entrained primary air theoretically required for complete combustion, percent, P = pressure of the fuel gas upstream of the entering orifice, d = specific

TABLE 13

**Values of a 10,000-Btu/hr Atmospheric
Burner's Air Injecting Ability for a High-
Methane Natural Gas and Hydrogen**

Fuel gas

	Natural gas	Hydrogen
P (in. wc)	7.0	7.0
d (air = 1)	0.6	0.07
H (Btu/SCF)	1,059	32.5
R (Btu/hr)	10,000	10,000
$Pd^{0.15}$	1.43	0.837
$H^{0.5}$	32.54	18.03
$R^{0.5}$	100.0	100.0
$IP/K \times 10^5$	43.9	46.4

TABLE 14

**Minimum Primary-Air Requirements for Various Types of
Appliance Burners**

Type of burner	Primary-air portion of the total theoretically required air (%)
Range top	55-60
Range oven	35-40
Water heater	35-40
Radiant-type space heater	65
Other heating appliances	As low as 35

From IGT Home Study Course — Gas Distribution, Institute of Gas Technology, Chicago, 1963. With permission.

gravity of the fuel gas, H = heating value of the fuel gas, R = energy input rate, and K = experimentally determined constant.

Table 13 gives the values calculated for comparison of the ability of an assumed 10,000 Btu/hr input rated atmospheric burner to entrain primary air using gases (natural gas and hydrogen) that have densities that differ by approximately 860%. The values of IP/K for hydrogen and for natural gas indicate that a higher percentage of primary air will be entrained by hydrogen than by natural gas.

Table 14 gives the typical primary-air requirements for appliance burners operating on natural gas. If we assume the range-top burner is adjusted for natural gas (60% primary air), and we replace the natural gas with hydrogen, the primary aeration for the substitution would equal, by Equation 30, 85%.

The burning velocity of natural gas, using 100% of the theoretically required air, is about 1 ft/sec.[12] The burning velocity of hydrogen, however, has been measured at 9.2 ft/sec using only 57% of the theoretically required air. Because the hydrogen flame is so much faster than the methane flame, there may be the difficulty of flashback with hydrogen burners.

Flashback is the propagation of the flame front back through the burner ports and upstream to the metering orifice. The flame cannot travel through the orifice because the gas upstream is pure fuel and cannot sustain combustion. However, if combustion

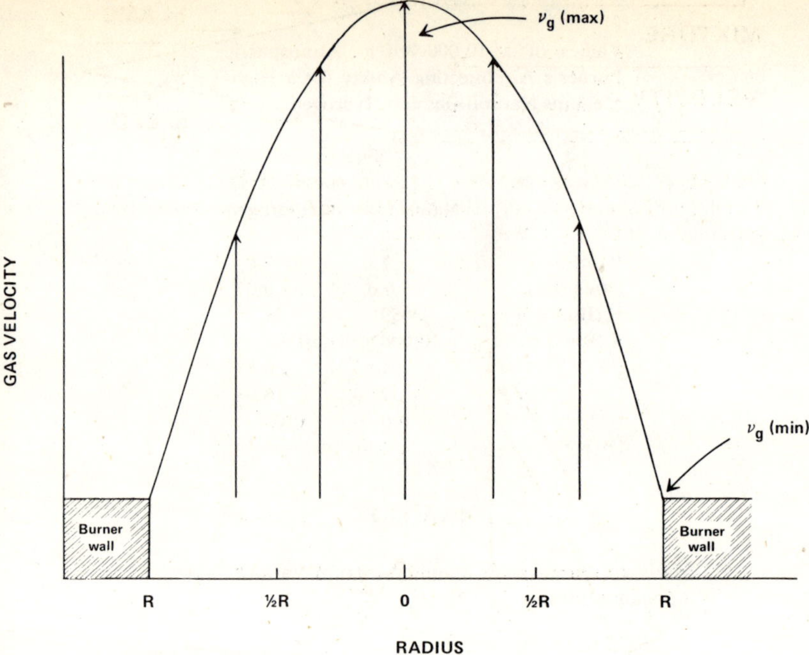

FIGURE 10.   Parabolic velocity profile of a stream at a burner port.

takes place at the metering orifice instead of at the burner ports, the burner head may
be severely damaged.

The tendency of a burner to flash back is indicated by the boundary velocity gradient
of the burner port. The boundary velocity gradient can be derived from Poiseuille's
Law (the basic equation for laminar flow in ducts), which is expressed as[23]

$$\nu_g = k(R^2 - r^2) \tag{29}$$

where ν_g = gas velocity, k = a constant, R = gas stream radius, and r = distance
from stream center. The value of the constant (k) can be determined by using volume
flow per unit time — i.e., gas flow rate — and the tube radius (R):

$$k = 2V/\pi R^4 \tag{30}$$

where V is the volumetric flow rate. Substituting the value of k into Equation 29, the
gas velocity (ν) is calculated to be

$$\nu_g = (2V/\pi R^4)(R^2 - r^2) \tag{31}$$

The average value of the gas velocity ($\bar{\nu}_g$) is

$$\bar{\nu}_g = 4V/3\pi R^2 \tag{32}$$

Figure 10 shows the velocity profile of a gas stream from a burner port. Figure 11
shows the interplay between the gas velocity (ν_g) and the flame speed (ν_f).

Assuming that the *inner* cone* of a flame from a Bunsen burner is a right cone —

---

*   It is also assumed that within the inner cone there exists only an unburned gas mixture.

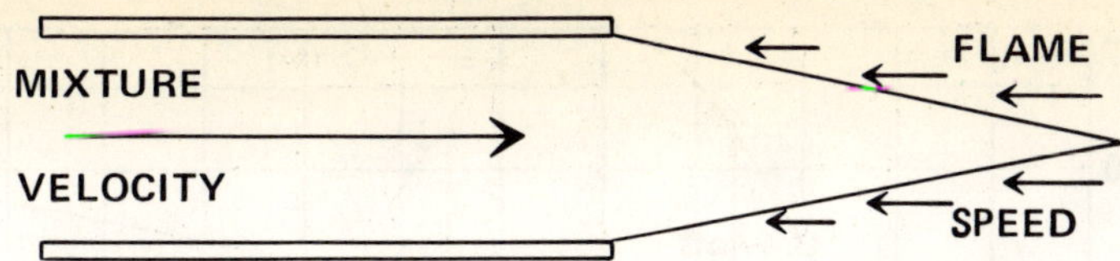

FIGURE 11.   Mixture velocity vs. flame speed. (From *Combustion Technology Manual,* 2nd ed., Industrial Heating Equipment Association, Washington, D.C., 1974. With permission.)

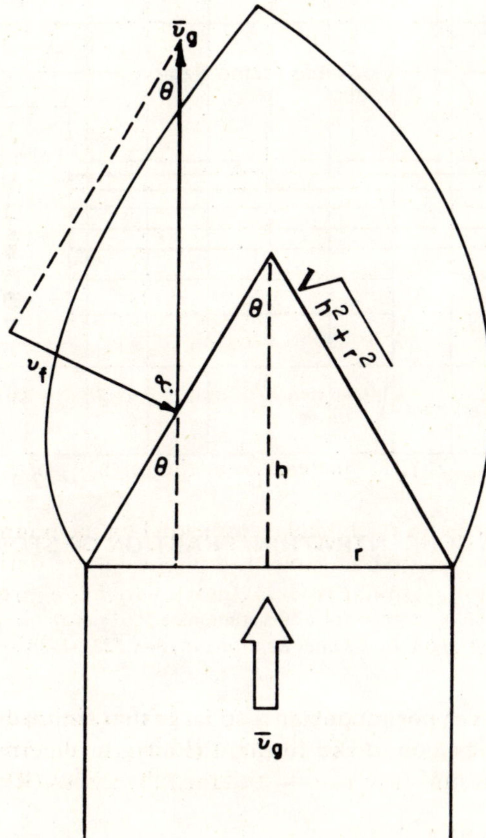

FIGURE 12.   Flame from a Bunsen burner with a true cone-of-revolution inner cone.

that is, a true cone of revolution — and that the air-gas flow is laminar, by definition,[12] the flame velocity (v_f) is the component of the average air-gas velocity (v_g) in a direction perpendicular to the flame front, as shown in Figure 12. Therefore, from Figure 12

$$v_f = \bar{v}_g \sin \theta \tag{33}$$

under stable conditions.

If Equation 29 (Poiseuille's Law) is differentiated with respect to r, setting r equal to R, the boundary velocity gradient can be obtained:

$$\text{grad } v_b = 4V/\pi R^3 \tag{34}$$

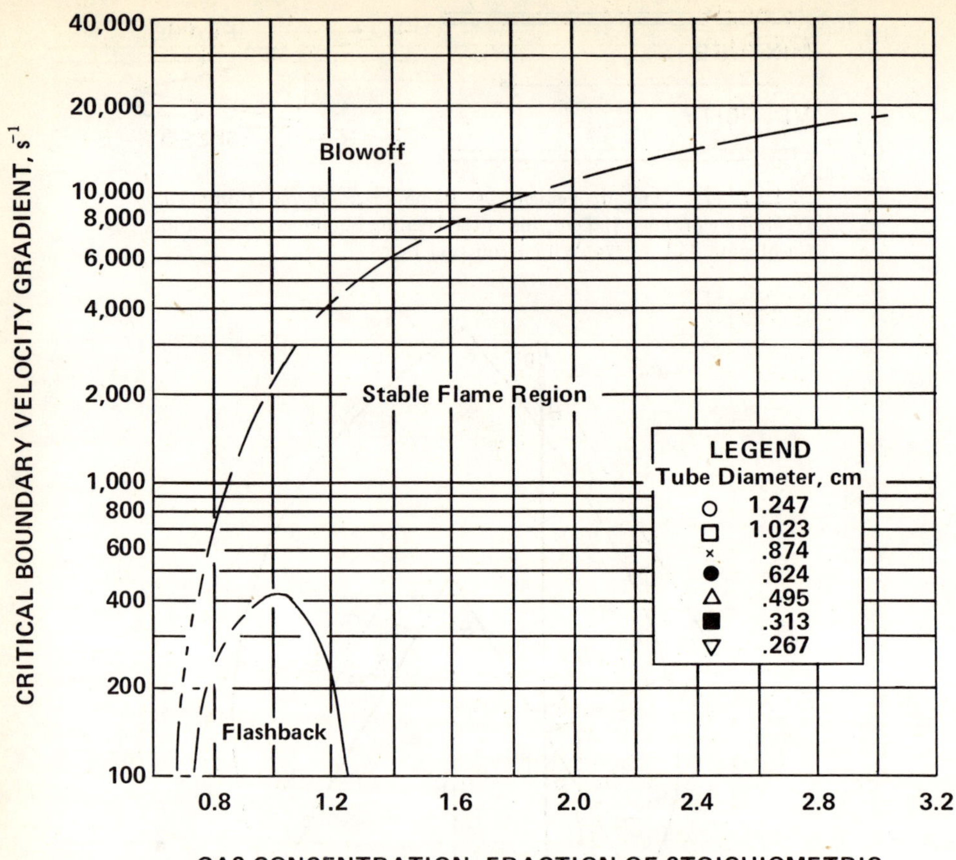

FIGURE 13.   Flame stability diagram for a fuel containing 100% methane. (From Grumer, J., Harris, M. E., and Rowe, V. E., *U.S. Bur. Mines Rep. Invest., No. 5225, 1956.*)

However, when the burner port diameter is so large that the air-gas velocity (v_g) profile at the wave fringe may be considered linear, the air-gas velocity equation may be approximated by the equation

$$v_g = \text{grad } v_b \cdot d \tag{35}$$

where d is the distance from the stream boundary.

Thus, the gas velocity at any d is proportional to the velocity gradient, and if a state is reached in which the gas velocity (v_g) at some point becomes smaller than the burning velocity ($\overline{v}_f$), the combustion wave propagates back against the gas stream into the tube — that is, it flashes back. At limiting conditions for lifting flames (blowoff), grad v_g becomes the critical boundary velocity gradient. At this value, the air-gas velocity (v_g) exceeds the burning velocity (flame speed, v_f), and blowoff occurs.

The results of the investigations by Grumer et al. (of the U.S. Bureau of Mines)[17] concerning flashback, lift-off, and stability of methane-air and hydrogen-air mixtures are shown in Figures 13 to 15. In general, a hydrogen flame can be leaner than a methane flame before blowoff will occur; however, the velocity gradient (v_b) must be maintained at a higher level to prevent flashback. To prevent flashback, the values from Figures 13 to 15 indicate that, at the critical boundary values for stoichiometric

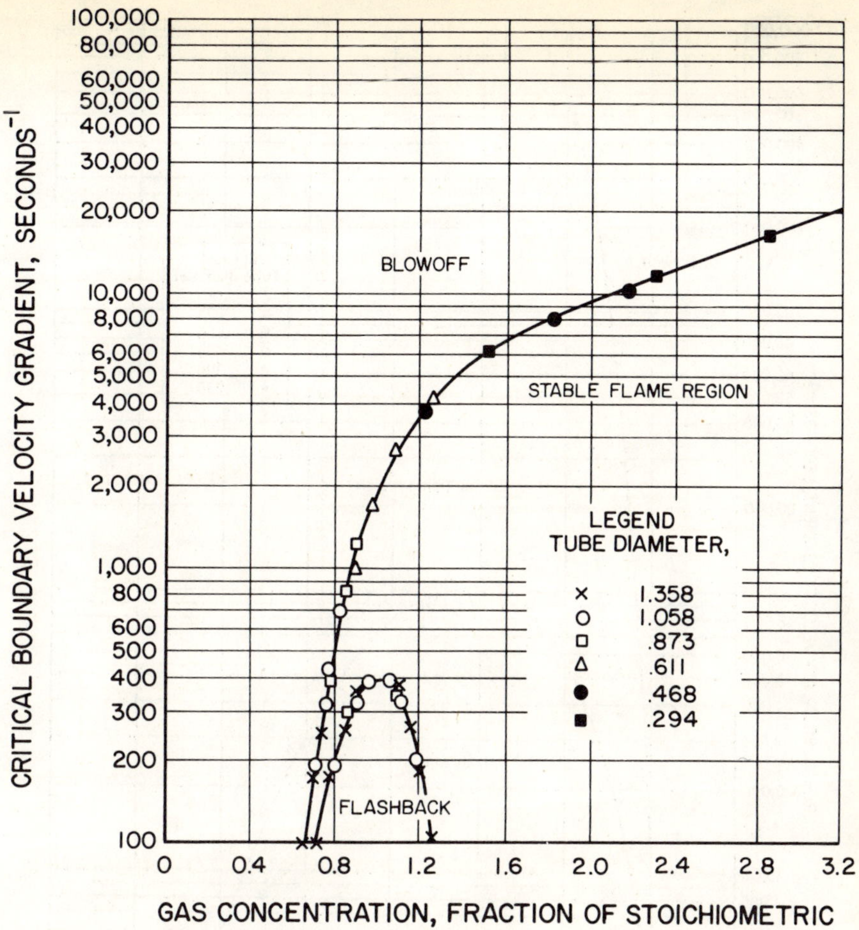

FIGURE 14.   Flame stability diagram for natural gas (91.5% CH_4, 5.2% C_2H_6, 1.3% C_2H_8, 0.9% CO_2, 0.6% N_2, 0.2% C_3H_6, 0.2% C_4H_{10}, 0.1% C_4H_8).

mixtures of hydrogen and air, the average gas velocity (v_g) of hydrogen should be approximately nine times greater than the average velocity of methane.

The experimental results shown in Figures 13 to 15 were obtained from measurements taken on single-port Bunsen burners and are not strictly applicable to multiport burners found in contemporary appliances. However, if Equation 34 is applied to a typical Bunsen burner (for example, burner No. 1 in Table 11), the boundary velocity gradient is found to be approximately 2000 (assuming 60% primary air stoichiometric). Figure 13 shows that the burner will then be operating in stable-flame range. If the same burner were operated on hydrogen (with no adjustments), the boundary velocity gradient, calculated by using the analysis of the air/fuel ratio, primary-air entrainment, and fuel flow rate shown above, would be about 2600 at a gas concentration of 1.18 times stoichiometric. Figure 15 shows that flashback would then occur. Again, it should be remembered that the flashback characteristics of real appliance burners cannot be accurately determined by analysis of single-port burners and that their tendency to flash back when operated on hydrogen must be determined *experimentally*.

One of the consequences of the high flame speed of hydrogen is a change in the shape of the flame. A standard natural gas flame is a cone about 1/8 in. high, but with hydrogen, the flame will be much shorter. To show this, we assume a natural gas

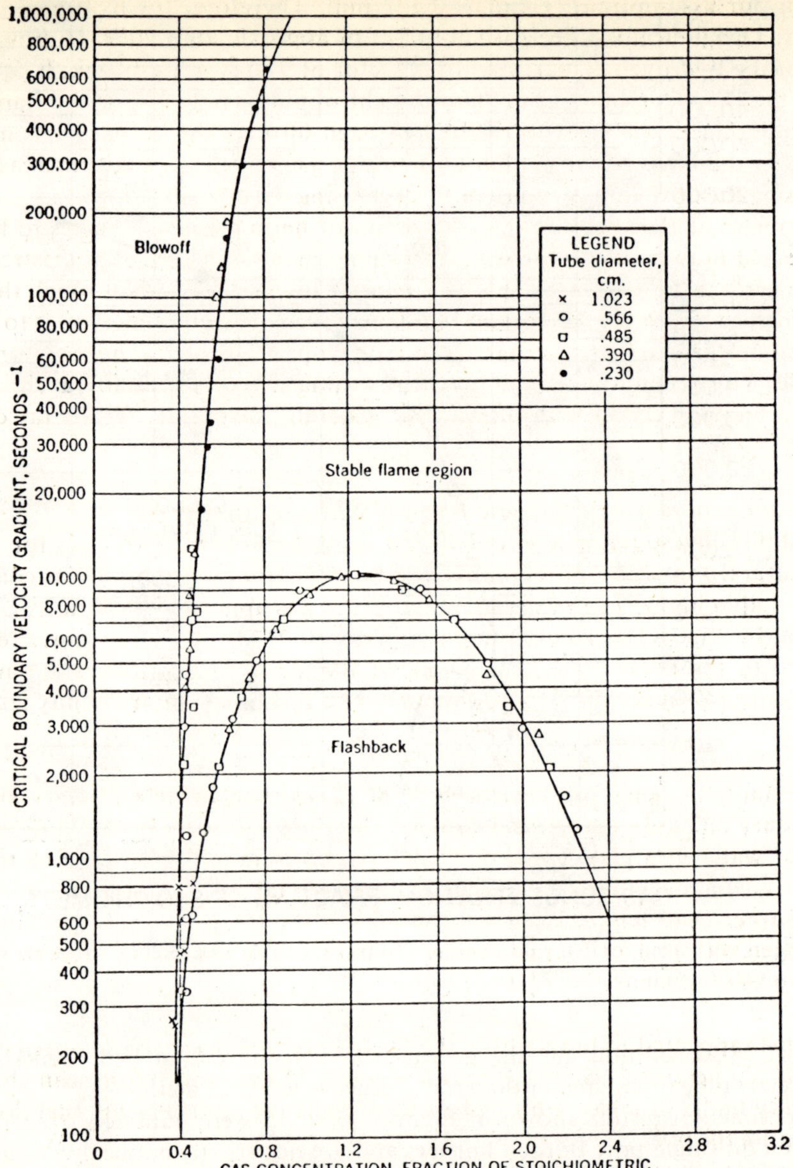

FIGURE 15.   Flame stability diagram for a fuel containing 99.7% hydrogen and 0.3% oxygen. (From Grumer, J., Harris, M. E., and Rowe, V. E., *U.S. Bur. Mines Rep. Invest.*, No. 5225, 1956.)

flame height of 3.17 mm and a port radius of 0.377 mm. From Equations 26 and 33

$$\nu_g \text{ (nat. gas)} = 8.5 \text{ ft/sec} \qquad (36)$$

$$\nu_g \text{ (hydrogen)} = (2.93)(8.5)\ 24.9 \text{ ft/sec}$$

From Figure 12

$$h \text{ (inner cone flame height)} = r \cot \theta \qquad (37)$$

and, from our assumptions, r equals 0.377 mm. Therefore, for hydrogen, h equals 0.948 mm. This indicates a decrease in height of approximately 70%. If, however, the velocity of the hydrogen is increased by a factor of 9 to stop flashback, h equals 3.11 mm; and the change in inner-cone flame height would be only 2%. If the flame height is to be maintained, the velocity of the hydrogen through the burner ports must be 9 times greater than that of the methane. (The port gas velocity could be increased either by increasing the upstream pressure or by decreasing the port diameters.)

The problem of flashback in the conversion of natural gas appliances to hydrogen can be treated in two ways. One way is to simultaneously increase the upstream pressure and decrease the port size. This can increase the flow velocity through the burner ports enough so that it is greater than the flame speed. The other method is to decrease the amount of primary air so that the mixture upstream of the burner ports is not flammable. This second technique has been evaluated experimentally at the Institute of Gas Technology (IGT) with mixed, but generally favorable, results (as discussed below).

3.5.1.2. Contemporary Atmospheric Burners Without Primary Air

A feasibility investigation was conducted using hydrogen as the fuel for a 12,000 Btu/hr (natural gas) atmospheric range-top burner. The burner head was constructed of stamped aluminum. The outlet pressure of the gas appliance regulator was factory adjusted to be 6 in. water column (approximately 0.217 lb/in.²). The only mechanical adjustment to the burner was the sealing of the air shutter so that the burner was unable to inject primary air by entrainment. The results of the feasibility study were that

1. The aluminum range-top burner rated at 12,000 Btu/hr (0% primary air, 100% secondary air) did operate in a cooking situation — i.e., it heated approximately 4 lb of water in a glass container (coffee pot) from approximately 68 to 212°F. Heat transfer and energy measurements were not taken.
2. The burner-head temperature was warmer to the touch of the investigator when hydrogen was used as the fuel than when natural gas was used; however, no quantitative measurements were taken.

The flames appeared to burn within the burner ports. Because flame speed is a function of gas temperature and pressure, an increase in gas temperature can change the flammability limits of a gas; with hydrogen, the change can cause the fuel (hydrogen) to burn within the ports.[12]

The calculated theoretical (adiabatic) flame temperature, for a stoichiometric hydrogen-air mixture and at an inlet gas temperature of 60°F, considering dissociation, is approximately 3525°F.[27] The flame temperature of a hydrogen-air mixture is, therefore, approximately only 10% higher than the flame temperature of a natural gas-air mixture, and this difference is probably not sufficient to produce a sensible difference. However, if *the specific flame intensities** of hydrogen[21] and natural gas[22] at stoichiometric mixtures are compared, where

---

* *Specific flame intensity* of a fuel gas can be defined as the *rate of heat release by the flame* of this gas when burned in a prescribed burner of definite design and at a definite inner-flame-cone height. This can be expressed mathematically as[12]

$$I = H\nu/A \qquad (38)$$

where I = the specific flame intensity (capacity) in Btu/sec-in.² port area, H = the *net* heating value in Btu/ft³ of gas-air mixture issuing from the burner in 1 sec, ν = the rate of flame propagation of the gas-air mixture in ft/sec, and A = the ratio of the burner area to the inner-flame-cone area.

TABLE 15

ANSI Carbon Monoxide Level Restrictions

Appliance	Amount of air-free carbon monoxide (ppm)
Ranges	800
Refrigerators	300
Others	400

$$I_{H_2} = 4.14 \text{ Btu/sec–in.}^2$$

and

$$I_{Nat.} = 0.646 \text{ Btu/sec–in.}^2$$

the specific flame intensity of the hydrogen-air mixture (I_{H_2}) is seen to be approximately *600%* greater than that of natural gas ($I_{Nat.}'$). For a gas-air mixture burned without primary air, the mathematical expression for flame intensity could be a qualitative indication of the flame to burner head heat transfer and of the heat available from the flame — thus possibly explaining why the burner head felt hotter to the investigator when the fuel was hydrogen.

Noise was generated upon burner start-up (ignition) and shutdown (extinction). The ignition noise was a sharp, cracking sound, whereas the extinction noise was a muffled sound of higher intensity than the ignition noise. On extinction of the burner, the noise was due to flashback occurring at a zero gas-flow rate. At ignition, the conditions favorable for the generation of noise were the ignition velocity (flame speed) of the hydrogen-air mixture and the composition of the mixture itself. There was no noise generated by the combustion waves *during* burner operation.

3.5.2. Emissions from Hydrogen-Fueled Burners

Currently, the American National Standards Institute (ANSI) has only one criterion for combustion pollutants from domestic gas appliances, and that is in regard to carbon monoxide. The maximum quantities of carbon monoxide allowed by ANSI are shown in Table 15.

Because hydrogen contains no carbon, its products of combustion contain no carbon monoxide or unburned hydrocarbons. However, laboratory investigators have shown[10] that an open-flame hydrogen burner produces significantly larger quantities of nitrogen oxides than an equivalent unit burning natural gas. Table 16 shows the nitrogen oxides emissions from hydrogen used in an atmospheric, open-flame range burner. It should be noted that natural gas used in the open-flame burner had nitrogen oxides emission levels of from approximately 80 to 100 ppm. From Table 16 it should also be noted that there was an increase of approximately 30% in nitrogen oxides emissions using secondary air only, compared with using hydrogen with primary air to fuel the open-flame range burner.

The most serious shortcomings of unmodified burners operated on hydrogen rather than natural gas, with respect to the criteria previously listed, will be flashback and noise. Another problem, not related to current burner requirements, will be the generation of nitrogen oxides.

There does not appear to be a simple way to change existing burner heads so that

TABLE 16

Nitrogen Oxides Emissions from an Open-Flame Range Burner Using Hydrogen as Fuel

	Air shutter	
	Fully open	Closed
Nitrogen oxides (ppm)	257	335
Increase (maximum) compared with natural gas (%)	221	319
Increase (minimum) compared with natural gas (%)	157	235

flashback will not be a problem. Burner ports can be enlarged by drilling (as they were when manufactured gas was replaced by natural gas), but reduction of the port size is not possible. The second way of overcoming flashback tendencies, which is by eliminating the primary-air supply, seems to lead to noise generation in burners (designed for use with natural gas), which have a large cavity upstream of the burner ports for gas-air mixing.

It would appear, then, that the easiest way to convert an existing natural gas appliance for operation with hydrogen would be to design a new burner head that could be installed in the field. Such a burner head could be designed to operate without primary air (the burner ports acting as the metering orifices), and the amount of gas available for "explosions" upon flame extinction could be minimized.

3.5.2.1. Burner-Head Port Sizing

A standard-sized burner head, 9000 Btu/hr, will pass about 9 ft^3 of natural gas and 47.3 ft^3 of air per hour. A replacement hydrogen burner head, operated without primary air, must pass 27.7 ft^3 of hydrogen to deliver heat at the same rate. Because the burner ports will be the metering orifices, the pressure upstream of the ports will be the distribution line pressure (about 7 in. water column). If we assume that eight ports are used per burner head (as in Figure 16), their size can be calculated from Bernoulli's theorem (Equation 25):

$$A = \frac{YC\,(h_L/\rho)^{0.5}}{q} \tag{39}$$

$$A = 0.000296 \text{ in.}^2$$

corresponding to a diameter of 0.0097 in.

3.5.2.2. Burner Configuration

Figure 16 is a suggested configuration for a hydrogen replacement burner for a range top or an automatic hot-water storage heater. An oven burner could be of long cylindrical design. However, because the anticipated small port areas may be prone to clogging by dust particles, a target or spider-type cover may be necessary to prevent blockage of the ports by foreign particles.

3.5.2.3. Burner Construction Material

Burner construction material would be determined by flame heating tests. Figure 17 shows one such method for determining temperature rise and heat transfer. The selection of an appropriate construction material would be based on (1) energy input, (2)

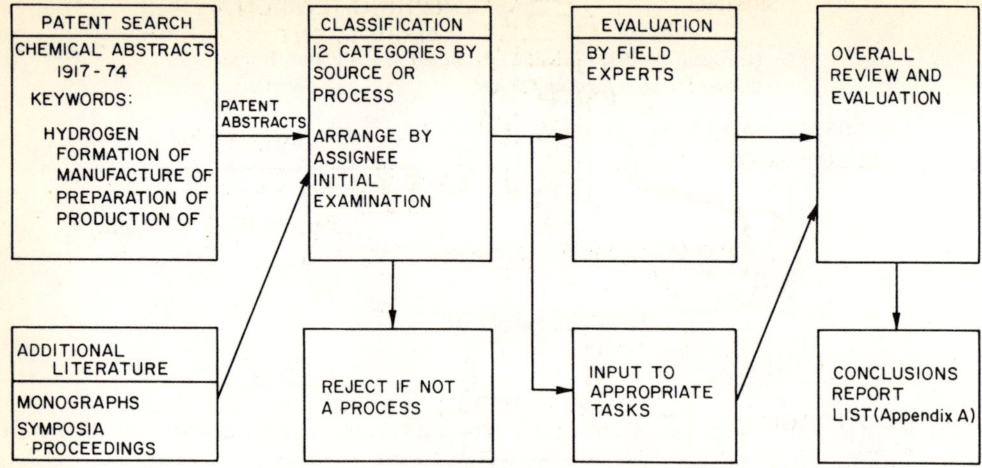

FIGURE 16.   Suggested hydrogen burner design.

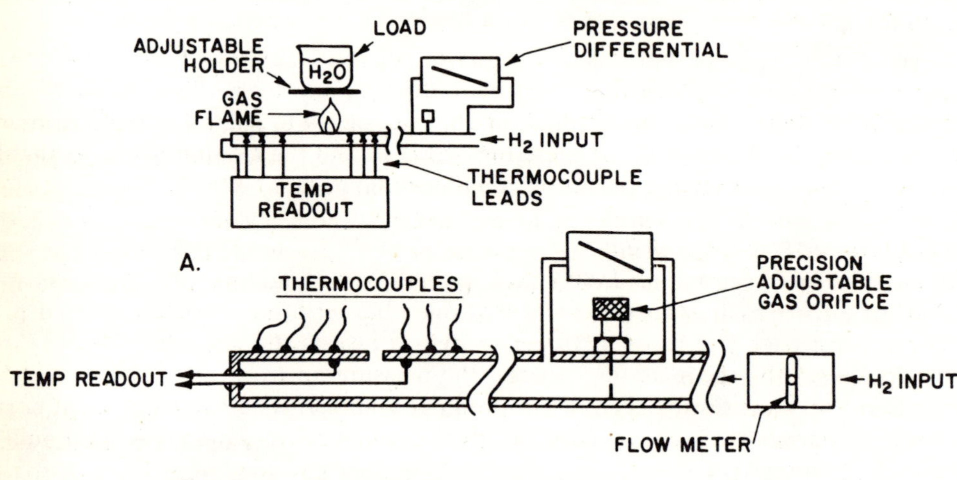

FIGURE 17.   Method for conducting hydrogen-flame heating tests.

pressure, (3) heat conduction, (4) heat radiation, (5) load distance from the burner, (6) heat reflection, (7) size, and (8) temperature rise.

3.5.2.4. Burner Ignition

The conventional gas pilot can be used to ignite the proposed hydrogen burner; however, it is more practical, in terms of fuel expenditure and air pollutant control, to use catalytic ignition. In catalytic ignition systems, the initial flow of gas (hydrogen) to the appliance passes over a catalyst. The catalyst causes the hydrogen to react with the oxygen in the air at room temperature and thereby produces heat. The catalyst is supported on a material that heats quickly during the catalytic combustion. When the surface temperature reaches the autoignition temperature of the hydrogen-air mixture (about 1085°F), a flame is initiated.

3.5.2.5. Noise

The ignition noise will not be completely eliminated because of the anticipated con-

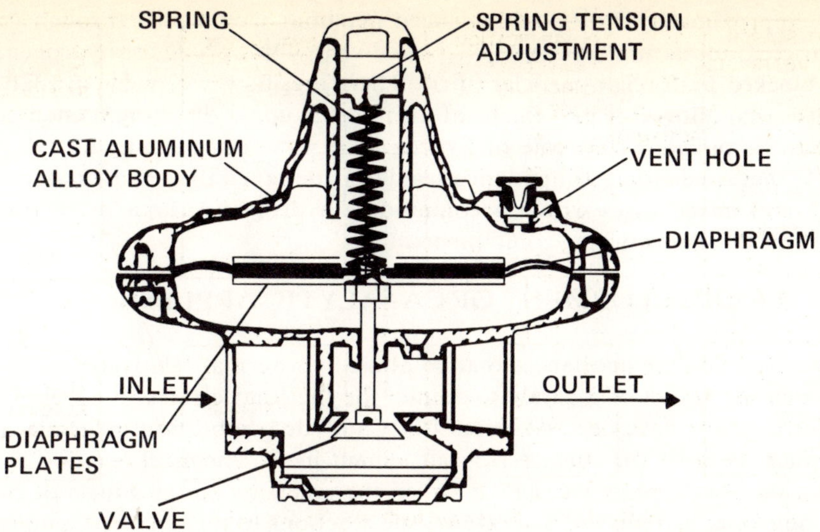

FIGURE 18.   Appliance regulator. (From *House Service Regulators*, and *Commercial and Industrial Service Regulators*, Rockwell Manufacturing Company, Pittsburgh, Pa. With permission.)

centration of hydrogen in the vicinity of the burner ports and the ignition. There should be a minimum amount of burner operation noise. The extinction noise generated by the hydrogen burner should be less than that generated by the contemporary range burner used in our feasibility investigation.* The reduction could be accomplished by a reduction in the available volume of hydrogen in the burner head at shutoff, i.e., the hydrogen burner would have less interval volume than the contemporary atmospheric gas appliance burner.

3.5.3. Appliance Regulators

Appliance regulators maintain a constant input pressure to the appliance, regardless of fluctuations in the supply pressure, and thus help to ensure optimum burner performance. In general, pressure regulators on domestic gas appliances use a vent for air movement from the atmospheric side of the regulator diaphragm. The working parts of such a domestic appliance regulator are shown in Figure 18.

ANSI Standard No. Z21.18-1969 (Standard for Gas-Appliance Pressure Regulators) regarding external leakage would apply to a regulator handling hydrogen fuel. The leakage performance test uses clean air as the gas. However, the area of the vent hole appropriate for natural, manufactured, mixed, and LP gas-air mixtures (specific gravity of approximately 0.64; air = 1) would be too large for use with hydrogen. The maximum allowable venting rate, as specified in ANSI Z21.18-1969 for fuel gases of specific gravities approximately equal to 0.64, is 2.5 ft³/hr.

The lower flammability limit for natural gas (at 1 at) is approximately 4.9%, by volume, in air.[14] For hydrogen, the lower flammability limit (at 1 at, 72°F) is approximately 4.0%, by volume, in air.[20] If we assume that the flow rate (2.5 ft³/hr) is such that, in a well-ventilated room, the lower explosion limit (LEL) for natural gas, i.e., 4.9% by volume, is not reached, the amount of hydrogen (specific gravity of 0.07; air = 1) that could be vented under these conditions would be approximately 2.0 ft³/hr.

---

*   The extinction noise generated by the contemporary natural gas burner operating on hydrogen was found to be objectionable.

However, approximately 7.0 ft³ of hydrogen per hour would pass through the vent sized (restricted) for natural gas. A smaller sized vent, appropriate for hydrogen, could easily be blocked by foreign particles (dirt). There are, however, vents available with check valves that allow free movement of air into the upper diaphragm chambers but that restrict the outward flow rate of hydrogen to some designated value. Informed opinion[16] is that residential gas appliance regulators designed to regulate natural, manufactured, and mixed gases would accommodate hydrogen without deterioration of the diaphragm or other working components.

3.6. DEVELOPMENT OF CATALYTIC APPLIANCES

It is possible to design appliances that combust hydrogen at relatively low temperatures through the action of a catalyst, despite the fact that efforts to develop similar burners for methane have been unsuccessful. Catalytic combustion takes place on an active surface. Because the surface is a participant in the chemical reaction sequence of combustion, the "energy barrier" between the reactants and products of combustion can sometimes be reduced. This allows the reactions to take place at temperatures below those characteristic of flames.

Research and development on catalytic appliances was conducted at Billings Energy Research Corp., Provo, Utah under the sponsorship of the Mountain Fuel Supply Co.,[15] and at the Institute of Gas Technology (IGT), under the sponsorship of Southern California Gas Co.[26] Billings Energy Research Corp. converted the cooking appliances in a Winnebago® motor home to catalytic hydrogen combustion. The research at IGT is concerned with the development of catalytic ranges, water heaters, and space heaters. Catalytic-igniter research, sponsored by several gas utility companies, is also under way at IGT.

There are two differing approaches to the design of hydrogen-fueled catalytic appliances. One is to design for pure catalytic combustion, wherein all combustion takes place through the action of a very active catalyst. The other is to use a flame to heat and assist a less active catalyst.

3.6.1. Low-Temperature Catalytic Appliances

When catalyzed by noble metals, hydrogen combustion in air can begin at room temperature. Heat is released as the hydrogen burns, thus raising the temperature of the catalyst and substrate; but combustion can be maintained at comparatively low (surface) temperatures. Most other *common* fuels must be heated to moderately high temperatures before they will *begin* to combust catalytically.

A low-temperature catalytic burner must maintain a steady rate of heat release to the load or to the environment so that it does not provide an ignition source for a flame. As was presented in the last section, it is quite possible to make catalytic igniters in which the combustion on the catalytic surface heats the catalyst to a temperature above the autoignition temperature of the mixture, thus initiating flame-type combustion. Stable catalytic combustion can be maintained if any of the three following criteria are met:

1. The laminar flow velocity of the hydrogen-air mixture over the catalyst is greater than the flame velocity of hydrogen. If this is true, a flame cannot propagate away from the catalyst. This approach has been tested experimentally and was found to be impractical.[26]
2. The composition of the hydrogen-air mixture passed over the catalyst places the mixture beyond the limits of flammability. This approach is impractical if the mix-

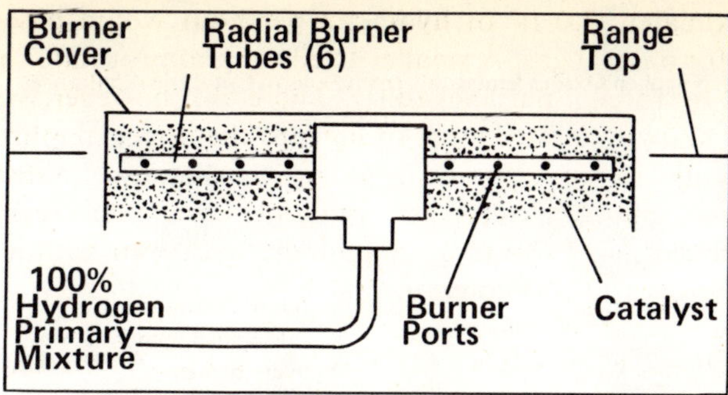

FIGURE 19.   Configuration of Billings Energy Research Corporation. Flame-assisted catalytic burner. (From Baker, N. R., paper presented at 9th Intersociety Energy Conversion and Engineering Conf., San Francisco, August 1974. With permission.)

ture is fuel rich because it implies that unburned hydrogen will be vented by the appliance. Results of experiments with mixtures that are less than 4% hydrogen (fuel lean) have not been promising.[15]

3.   No points on the surface of the catalyst ever reach the autoignition temperature of hydrogen-air mixtures (1085°F). This implies that a balance between the heat transferred away from the surface and the heat released by the combustion of incoming hydrogen is struck at a lower temperature.

Experiments at IGT have shown the third method to be the most practical.[26] Stable and complete catalytic combustion has been maintained in burner configurations at temperatures as low as 400°F (the surface temperature as determined by thermocouple measurements).

3.6.2. High-Temperature Catalytic Burners

At elevated temperatures, some materials that are not active catalysts for hydrogen combustion at low temperatures become active. Examples of such materials are iron and steel. Exploitation of this property by Billings Energy Research Corp. has resulted in a hybrid, flame-assisted catalytic burner. The configuration of this burner is shown in Figure 19.

During operation, the flame-assisted catalytic burner must be activated by an outside ignition source, such as a glow coil or a pilot light. At first, all combustion takes place in the flame, but as the flame heats the catalytic surface, proportionately more and more combustion takes place through the action of the catalyst.

3.6.3. Advantages of Catalytic Combustion

It should be noted that the designing of catalytic burners is not an exact science, and a great deal of development work must be done before such devices can be marketed. The incentives for undertaking such programs are clear from the results of the first experiments performed on early burner models. Catalytic appliances can be more efficient than conventional gas flame appliances and can virtually reduce exhaust emissions other than water vapor to zero.

When catalytic combustion takes place at temperatures below 1500°F, the formation of nitrogen oxides is, for all practical purposes, eliminated. Because no carbon monoxide is formed during hydrogen-air combustion, the only product of combustion is water vapor. Laboratory results of nitrogen oxides emissions from two low-tempera-

TABLE 17

Nitrogen Oxides Emissions from Several Catalytic Appliances

	Burner type	Nitrogen ox- ides (ppm)
Low-temperature catalytic[26,a]	Chimney	0.08
	Vertical fin	0.03
High-temperature catalytic[15]	Standard Al range top	5.2
	Stainless steel, experimental	1.8
	Al oven/broiler	4.5
	Cast iron	4.0

[a] Reported on an air-free basis.

ture catalytic appliances and from four high-temperature catalytic burners are shown in Table 17.

The data for low-temperature catalytic appliances are reported on an air-free basis, as were the measurements reported in Table 16. This means that the measured concentrations are adjusted to exclude the dilution effect of excess air. With excess air, the nitrogen oxides concentrations would be even lower. The experimental apparatus for making such measurements is shown in Figure 20. Because Reference 15 does not specify whether or not the quoted results are on an air-free basis, it is assumed that they are measured concentrations that are diluted by an undetermined amount of excess air.

The nitrogen oxides production levels reported for the low-temperature catalytic space heaters are on about the same order of magnitude as the nitrogen oxides levels found in ambient air. Thus, it should be possible to build space heaters and other appliances that are ventless. In present-day furnaces, about 30% of the chemical energy in the fuel is lost when the combustion products are vented through chimneys. Therefore, ventless appliances could potentially operate much more efficiently than present-day appliances. At IGT, under certain intermittent operating conditions, second generation, experimental catalytic water heaters have had measured efficiencies of greater than 80%, based on that portion of the high heating value of hydrogen that is transferred to the water.

Preliminary calculations indicate that the water formed as a combustion product in ventless space heating may or may not be a problem, depending on the tolerable humidity level and the frequency of air changes in specific houses.[26] There is also the possibility with some appliances (water heaters, for example) of installing condensing units to trap combustion-produced water vapor. These traps would then permit full utilization, by the appliance, of the higher heating value of the hydrogen burned, and they would provide a supply of relatively pure water.

3.6.4. Cost of Catalytic Appliances

The cost of high-temperature catalytic appliances should not be significantly higher than that of standard appliances because the only addition is a stainless steel pad. Although low-temperature catalytic appliances make use of expensive metals (such as platinum), the catalyst-loading levels are so low that the appliance price should not increase significantly. (Preliminary estimates are that only about 10¢/1000 Btu-hr would be attributable to the catalyst material.) From Table 5, the estimated *peak-day* natural gas requirements for a 1200-ft² house is about 1.5 million Btu. The release of this much energy by the catalytic combustion of hydrogen would then require an in-

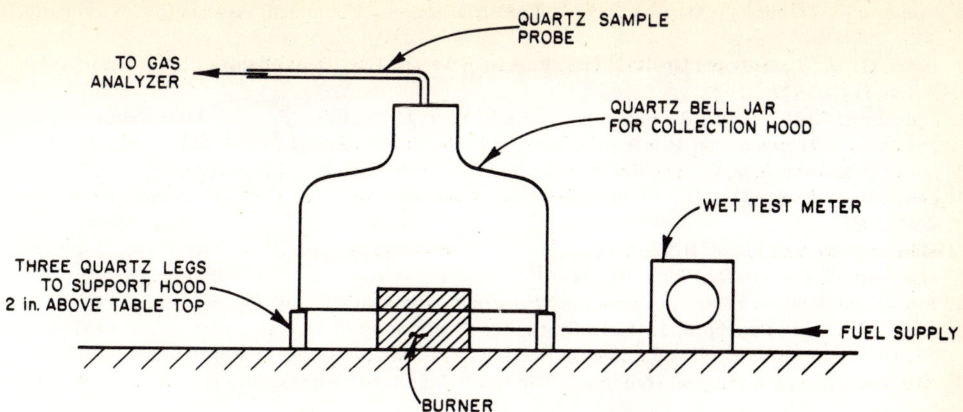

FIGURE 20. Burner-emission-testing setup.

vestment of only $12, attributable to platinum catalyst material. This figure is conservative (high) because ventless catalytic burners will be more energy efficient than standard burners and because low-temperature appliances will not need pilot lights.

REFERENCES

1. Gregory, D. P., Anderson, P. J., Dufour, R. J., Elkins, R. H., Escher, W. J. D., Foster, R. B., Long, G. N., Wurm, J., and Yie, G. G., A Hydrogen-Energy System, Catalog No. L21173, American Gas Association, Arlington, Va., August 1973.
2. 1973 Gas Facts, Catalog No. F10173, American Gas Association, Department of Statistics, Arlington, Va., 1973.
3. IGT Home Study Course — Gas Distribution, Institute of Gas Technology, Chicago, 1963.
4. Hale, D., *Pipeline Gas J.*, 201, 26, 1974.
5. Burnham, C. H., Freuchtenicht, H. L., Pflug, D. R., and Sweningsen, G. A., Transmission pipelines, in *Gas Engineers Handbook,* Industrial Press, New York, 1969, 818.
6. Wilson, G. G., Flow calculations and sizing of mains, in *IGT Home Study Course — Gas Distribution,* Institute of Gas Technology, Chicago, 1963, 62.
7. Flow of Fluids Through Valves, Fittings, and Pipe, Tech. Paper No. 410, Crane Co., Chicago, 1969.
8. Kuhlman, H. W., Wolter, F., Sowell, S., Beatty, G., Leininger, R. I., and McClure, G. M., The Development of Improved Plastic Pipe for Gas Distribution Purposes, American Gas Association, Arlington, Va., 1968.
9. Part 192, Til 6 49, Code of Federal Regulations, Subpart J, requirement 192,511, *Pipeline Gas J.*, 202, 49, 1975.
10. American Gas Association, Cleveland Laboratory, private communication, January 1975.
11. American National Standard for Household Cooling Gas Appliances, American National Standard Institute Z21 1-1974, American Gas Association, Arlington, Va., 1974.
12. American Gas Association, *Gas Engineers Handbook,* Industrial Press, New York, 1969.
13. Influence of Port Design and Gas Composition on Flame Characteristics of Atmospheric Burners, Res. Bull. No. 77, American Gas Association Laboratories, Arlington, Va., 1958.
14. *Gaseous Fuel — Properties, Behavior, and Utilization,* 2nd ed., American Gas Association, New York, 1954.
15. Baker, N. R., Oxide of nitrogen control techniques for appliance conversion to hydrogen fuel, paper presented at 9th Intersociety Energy Conversion and Engineering Conf., San Francisco, August 1974.
16. Dufour, R. J., consultant to IGT.
17. Grumer, J., Harris, M. E., and Rowe, V. E., Fundamental flashback, blowoff, and yellow-tip limits of fuel gas-air mixtures, *U.S. Bur. Mines Rep. Invest.,* No. 5225, 1956.
18. *Residential Energy Consumption,* Hittman Associates, Inc., Columbia, Md.

19. *Combustion Technology Manual,* 2nd ed., Industrial Heating Equipment Association, Washington, D.C., 1974.
20. **Jones, G. W.,** Tech. Paper No. 450, U.S. Bureau of Mines, U.S. Department of the Interior, Washington, D.C., 1929.
21. **Ladenburg, R. W., Lewis, B., Pease, R. N., and Tayler, H. S., Eds.,** *Physical Measurements in Gas Dynamics and Combustion,* Princeton University Press, Princeton, N.J., 1954, 409.
22. **Lewis, B. and von Elbe, G.,** Stability and structure of burner flames, *J. Chem. Phys.,* 11, 75, 1943.
23. **Lewis, B. and von Elbe, G.,** *Combustion Flames and Explosions of Gases,* Academic Press, New York, 1951.
24. **Mason, D. M. and Eakin, B. E.,** Calculation of heating value and specific gravity of fuel gases, *Inst. Gas Technol. Chicago Res. Bull.,* No. 32, 1961.
25. *Residential-Appliance Gas Consumption,* Northern Natural Gas Co., Omaha, September 1970.
26. **Sharer, J. C. and Pangborn, J. B.,** Utilization of hydrogen as an appliance fuel, paper presented at The Hydrogen Economy Miami Energy (THEME) Conf., Miami Beach, March 18 to 20, 1974.
27. *North American Combustion Handbook,* The North American Manufacturing Co., Cleveland, 1952.

Industrial Applications of Hydrogen

Chapter 4

INDUSTRIAL APPLICATIONS OF HYDROGEN

D. Cooperberg

TABLE OF CONTENTS

INTRODUCTION

Other chapters of this text will cover utilization of hydrogen as an energy mover. This chapter will attempt to review the current industrial usages of hydrogen as well as highlight potential areas where such industrial usage could become significant in the future.

4.1. CURRENT DAY USAGE

Today the largest industrial application of hydrogen is as a chemical feedstock for production of certain chemicals. As shown in Table 1, approximately 95% of all hydrogen consumed for industrial purposes is consumed as a chemical feedstock. The bulk of chemical hydrogen consumption is in one of three main industries: ammonia manufacture, petroleum refining and petrochemicals, and methanol production. In

TABLE 1

Recent U.S. Hydrogen Consumption (10^9 m^3)

End use	1968	1971	1972	1973
Ammonia	24.5	28.3	30.7	31.3
Refinery operations	22.2	37.8	38.8	40.6
Methanol	4.11	5.41	6.4	8.58
Hydrogenation of oils	0.31	0.34	0.40	0.42
Miscellaneous[a]	7.25	4.93	5.27	5.69
Total	58.3	76.8	81.6	86.7

[a] Electronics and electrical, space, other industrial, and other chemicals, for example, hydrogen peroxide, alcohols.

From Kelley, J. H. and Tasimann, E. A., Hydrogen Tomorrow...Demands and Technology Requirements, report of the Hydrogen Energy Systems Technology (HEST) Study for NASA, Jet Propulsion Laboratory Report JPL 5040-1, December 1975.

TABLE 2

Ammonia Usage

Direct usage (as liquid fertilizer)
Caprolactam production[a]
Acrylonitrile production[a]
Urea Production
   Used as fertilizer
   Used for resins manufacture (e.g., TDI)
Nitric acid production
   Used in fertilizer production (e.g., NH_4NO_3)
   Production of explosives
   Aniline production

[a] Utilized in fibers industry.

addition to these areas, minor quantities of hydrogen are consumed in metals treating, edible oil production, and other applications. The following text will review hydrogen utilization within each industry.

4.1.1. Ammonia Production

As shown in Table 1, ammonia production represents the largest consumption of hydrogen in the industrial sector today. It is used primarily for the production of fertilizer-based products as well as other products used in the fiber industry and other chemical processing industries. Table 2 shows some of the major usages for ammonia.

Ammonia is produced by the Haber-Bosch process developed in 1913. In this process, a mixture of nitrogen and hydrogen in a 1:3 ratio is passed over an iron-oxide promoted catalyst to produce ammonia by the following reaction:

$$3H_2 + N_2 \rightarrow 2NH_3$$

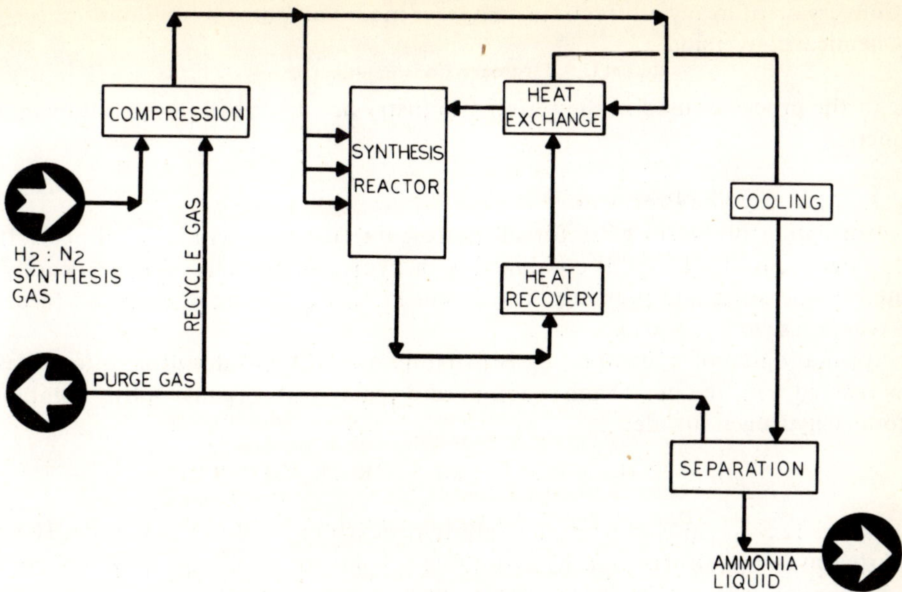

FIGURE 1.   Ammonia synthesis.

The reaction is exothermic, and typical reaction conditions range from 750 to 950°F and from 2000 to 5000 psig.

Figure 1 is a simplified schematic of ammonia synthesis process. In this process, the synthesis gas (H_2/N_2 mixture) is mixed with a recycle stream and subsequently passes to the ammonia converter. Reactor effluent is subsequently cooled, and ammonia product is recovered. The balance of the reactor effluent is recycled to mix with the fresh feed. A portion of the reactor effluent is purged from the synthesis loop in order to remove any inerts (e.g., methane) that may have been introduced in the synthesis gas.

Hydrogen for the production of the ammonia synthesis gas (i.e., H_2/N_2 in 3:1 ratio) is produced primarily by the reforming of natural gas. In the U.S. this method probably accounts for 90 to 95% of the ammonia synthesis gas produced. The production of synthesis gas by natural gas reforming is very similar to the production of hydrogen from natural gas as described in Volume I, Chapter 3. The primary difference is the inclusion of a secondary reforming step when producing ammonia synthesis gas. The reaction step allows for the addition of air-supplied nitrogen to a hydrogen-rich stream. The overall reaction for this secondary reforming may be represented as follows:

$$CH_4 + 2O_2 \rightarrow CO_2 + 2H_2O$$

$$2CH_4 + 2H_2O \rightarrow 6H_2 + 2CO$$

4.1.2. Petroleum Refining

Hydrogen is consumed in the petroleum refining industry in several processing options. The total hydrogen consumed by these operations makes the petroleum refining industry the largest consumer of hydrogen today (see Table 1).

In the petroleum refining industry, hydrogen is used primarily to achieve one of three processing goals:

1.   Removal of sulfur from refinery products

2.   Conversion of heavy oil fractions to lighter more valuable oil fractions
3.   Chemical conversion

Some of the processes used in the refining industry are described in the following text, by function.

4.1.2.1. Hydrotreating Processes

Hydrotreating processes exist for all petroleum fractions within a refinery. In all cases, a petroleum feedstock is combined with hydrogen and passed over a catalyst at elevated temperature and pressure. The extent of hydrotreating depends on feedstock type, reaction severity, and catalyst.

The primary goal of hydrotreating is usuaully one of hydrodesulfurization. Hydrogen is reacted with the petroleum fraction and reacts with organic sulfur compounds to produce hydrogen sulfide:

$$H_2 + HS(R) + R'SR''_2 \rightarrow 2H_2S + RH + R'H + R''H$$

The gaseous H_2S is removed from the light hydrocarbon gases (i.e., H_2 and CH_4) present during hydrodesulfurization by a suitable solvent solution (e.g., monoethanolamine). By similar reaction, hydrogen can react with nitrogen-containing compounds to form ammonia which is subsequently removed. For sufficiently severe hydrotreating conditions, hydrogenation of unsaturated hydrocarbons may also occur.

Severity — Low severity operations are usually associated with hydrotreating of naphtha fractions. In such operations, hydrogen consumption may range from 50 to 200 scf (H_2) per barrel of feed. Besides desulfurization, such treatment provides improvement in color and stability. High severity operations are associated with hydrodesulfurization of heavy petroleum fractions high in sulfur content (e.g., hydrodesulfurization of vacuum gas oil and atmospheric residua) and may result in a hydrogen consumption in the range of 600 to 1000 scf (H_2) per barrel of feed.

Usage — Hydrotreating in refinery operations can be used to improve a petroleum fraction qualities prior to further processing or to improve the qualities of a finished product. Processes for which hydrotreating is used to improve feedstock quality include catalytic reforming and catalytic cracking. Hydrotreating of catalytic reformer feedstock removes sulfur, nitrogen, and metal compounds that could subsequently poison a reforming catalyst. Similar benefits occur when treating catalytic cracking feedstocks. Hydrotreating of finished products is used to sweeten these products (i.e., remove sulfur compounds), improve stability (by hydrogenating certain unsaturates), and improve their color properties. Hydrotreating is also growing in use as more and more olefins projects are built. Olefins plants, which primarily produce ethylene and propylene, will also produce other liquid products which may contain acetylene or diolefins material. These compounds tend to polymerize and form sludge upon standing in storage. Hydrotreating is used to hydrogenate such compounds, converting them to more stable paraffin compounds.

4.1.2.2. Hydrocracking

Hydrocracking operations within a refinery are used to convert heavier, less profitable liquid fractions into lighter, more profitable fractions. An example of this is the conversion of vacuum gas oils and vacuum residium into lighter distillate fractions. Hydrocracking is basically a higher severity version of hydrotreating. In such reactions, pressures may range from 1500 to 3000 psig as compared to the lower pressures of hydrotreating.

In hydrocracking, hydrogen is used to saturate the cracked products produced under the severe operating conditions of the hydrocracking operations. In addition to cracking and saturation of hydrocarbons, the hydrogen simultaneously desulfurizes the hydrocarbon feedstock. Hydrocracking is thereby able to produce a low sulfur, lighter product from heavy hydrocarbons feeds.

4.1.2.3. Hydrodealkylation

Hydrodealkylation is the selective hydrogenation of aromatic compounds that leaves the benzyl or other aromatic ring structure unaltered. Produced in this way are high purity benzene, toluene, and naphthalene. For the production of benzene, a feed (toluene, xylenes, C_9, and aromatics) together with a hydrogen-containing stream are heated to the reaction temperature and passed over a dealkylation catalyst. Benzene and heavier aromatics are condensed, and lighter vapors may be purified and/or recycled. Benzene is purified in a benzene fractionator to produce the desired specification benzene. Other hydrogen consuming chemical conversions are the saturation of aromatics to produce high quality jet fuels as well as high purity hexane from benzene.

4.1.3. Methanol Production

Methanol production is the third largest user of hydrogen. Methanol is one of the major raw materials of the organic chemical industry. It is used as a feedstock for the production of formaldehyde resins (and subsequent wood laminates), dimethyl terephthalate (and subsequent polyester fibers), methacrylates, methylamines, acetic acid, and solvents. Methanol may also be used in future applications as a liquid fuel or as a fuel additive to extend conventional fuels.

Methanol is produced by first generating a synthesis gas suitable for methanol synthesis. Each of the following hydrocarbons can be utilized for production of methanol synthesis gas: natural gas, naphtha, and coal or heavy oils.

Natural gas feed — As with ammonia manufacture, steam reforming of natural gas is the most common method for supplying hydrogen for methanol synthesis. The steam reforming of natural gas produces a synthesis gas with an H_2 to CO ratio in excess of that required for methanol synthesis. Quite often, methanol plants will be located near ammonia plants to take advantage of a cheap source of carbon dioxide. The excess hydrogen can be reacted with make-up carbon dioxide to produce methanol by the following reaction:

$$3H_2 + CO_2 \rightarrow CH_3OH + H_2O$$

Naphtha feed — If naphtha feed is used for preparation of methanol synthesis gas, the hydrogen to carbon oxides ratio as defined below is usually in balance:

$$H: [CO + 1.5\, CO_2]$$

Coal and heavy oil — These hydrocarbon sources can also be used to produce methanol synthesis gas. Partial oxidation is the process most suitable for converting heavy oils and/or coal into methanol synthesis gas. However, the gasification product from this process lacks sufficient quantities of H_2 to convert all the carbon oxides into methanol. Hydrogen deficiency is due to lack of hydrogen in the feedstock. Additional hydrogen is generated by reacting steam with the gasifiers effluent in a water-gas shift reactor to accomplish the following reaction:

$$H_2O + CO \rightarrow H_2 + CO_2$$

TABLE 3

Typical Industrial Hydrogen Requirements

Use	m^3 of H_2	Product (unit)
Ammonia synthesis	2.18—2.50	NH_3 (1kg)
Methanol synthesis	1.62—2.34	CH_3OH (1kg)
Ethylene glycol[a]	2.6	Ethylene glycol (1kg)
Petroleum refining (av)	109	To refine 1m³ of crude oil
Hydrotreating		
Naphtha	9	Naptha (1m³)
Coking distillate	134	Distillate (1m³)
Hydrocracking	267—623	Feedstock (1m³)
Coal conversion to[a]		
Liquid fuels	784—1514	"Synthetic" oil (1m³)
Gaseous fuel	1.57	"Synthetic" gas (1m³)
Oil shale conversion to[a]		
Liquid fuel	232	"Synthetic" oil (1m³)
Gaseous fuel	1.21	"Synthetic" gas (1m³)
Liquid fuel from tar sands	303	"Synthetic" oil (1m³)
Iron ore reduction[a]	0.624	Iron (1kg)
Hydrogenation of fats and oils	0.109	Product (1kg)

[a] Future markets.

From Kelley, J. H. and Tasimann, E. A., Hydrogen Tomorrow… Demands and Technology Requirements, report of the Hydrogen Energy Systems Technology (HEST) Study for NASA, Jet Propulsion Laboratory Report JPL 5040-1, December 1975.

Sufficient carbon dioxide is removed from the shift reactor effluent to yield a synthesis gas suitable for methanol production.

Methanol synthesis — Hydrogen and carbon oxides synthesis gas, free of potential catalyst poison such as sulfur and chlorine, are then added to the methanol loop. Methanol synthesis is carried out over a catalyst at elevated temperature and pressure. Recent trends have seen the development of active methanol catalysts that have allowed synthesis reactor pressure to be reduced from a range of 3000 to 5000 psig to 700 psig. In the current environment, the methanol plant for chemicals productions will operate at low pressures, 750 to 1000 psig. In future methanol applications, this trend may reverse if large-tonnage methanol plants are required. This is due to the potential for building methanol plants for fuels production. Such plants would be considerably larger than the chemicals plants, and higher pressure synthesis process may be utilized to minimize this size of equipment required for large plants.

4.1.4. Miscellaneous Applications

Hydrogen finds application in many other industires, from food processors to electronics to glass. The consumption, however, is low in comparison to the previously discussed industries (see Table 3).

Unsaturated organic oils of soybeans, fish, cotton seed, corn, peanut, and coconuts are hydrogenated. This hydrogen treatment stabilizes these oils and improves odor, color, and consistency characteristics. Inedible oils and greases are hydrogenated for the production of soap and animal feed. Lubricants for use in food-processing equipment are produced by hydrogenation and purification of various greases to rigid specifications.

Much chemical synthesis not directly related to petrochemicals requires hydrogen. Hydrogen peroxide, hydrogen chloride gas, uranium hydride, and many organic chemicals all use hydrogen. The pharmaceutical industry uses hydrogen extensively to manufacture certain drugs.

Hydrogen is consumed in the oxy-hydrogen cutting of glass and welding of metals. It is also used in working leaded glass where the hydrogen content must be carefully controlled to prevent burning the lead out of the glass.

4.2. POTENTIAL FUTURE USAGE

Some of the processes described in the following text have already been developed on a commercial scale. However, the usage is still insignificant when compared to the potential usage in the future. Some of the more important processes that may be large consumers of hydrogen in the future are described in the following text.

4.2.1. Direct Reduction of Metals

Though currently small in hydrogen consumption, a potentially large market for future usage of hydrogen is the direct reduction of metal ores to the base metal. In such processes, hydrogen or a mixture of hydrogen and carbon monoxide gas (called reducing gas) is reacted with the metallic ore. Usually, the reducing gas flows upward through a falling or fluidized bed of metallic ore.

The iron industry has built several processing plants in which iron ore is reduced by hydrogen or a mixture of hydrogen and carbon monoxide. Reaction temperatures range from 550 to 780°C, and pressures range from 2 to 15 atm. In these processes, finely divided iron ore particles are contacted with the gas with the following reactions occurring:

$$3H_2 + Fe_2O_3 \rightarrow 2Fe + 3H_2O$$

$$3CO + Fe_2O_3 \rightarrow 2Fe + 3CO_2$$

In these processes, approximately 85 to 90% of the ore can be metallized. The ore is subsequently briquetted for shipment and storage or used directly in steel-making processes.

A strong incentive for this process has been the development of iron ore reserves in those parts of the world where coal is not readily accessible or where natural gas is relatively cheap. With cheap natural gas, a reducing gas mixture rich in hydrogen can easily be generated by the steam-methane reforming reaction. Nuclear power for generation of electricity to be used in producing hydrogen by electrolysis may also be considered where iron ore is plentiful but carbon resources are not. Wide acceptance of this technology depends on the economics of hydrogen and on the solution to technical difficulties (agglomeration and product phyrophoricity). Hydrogen reduction for nonferrous metallurgy (copper, zinc, uranium, and lead) is equally feasible but would be less hydrogen consuming. Direct reduction processes offer several advantages such as flexibility of plant location, reduced use of fossil fuels, and reduced pollution problems.

4.2.2. Coal Conversion

In the future, if conversion of coal to synthetic natural gas (SNG) or to synthetic petroleum liquids proves feasible, massive quantities of hydrogen will be required. The reason is quite simple when one looks at the chemical nature of coal. Coal has a carbon to hydrogen weight ratio in the order of 25:30. However, synthetic natural gas has a carbon to hydrogen weight ratio of 3, and synthetic petroleum liquids have a ratio of

8:10. It is obvious that to achieve these ratios, hydrogen must be added to the coal, or carbon must be removed. Most technologies being piloted are based upon hydrogen addition.

4.2.2.1. Gasification for Synthetic Natural Gas (SNG) Production

Today many gasification concepts are being evaluated in pilot plant, demonstration unit, and full-scale production plants. In all processes, the coal is gasified with the intent of producing a maximum amount of methane within the gasification reactor. To do this, reactor pressures are kept as high as possible, usually being limited by the ability to feed coal to the reactor and/or discharge ash from the reactor. The commercially available Lurgi process operates at approximately 30 atm, while the new processes (such as the Synthane process being developed by the Energy Research and Development Administration) operate at 65 atm. Unfortunately, no one-step gasifier exists that will convert all of the coal directly to synthetic natural gas (i.e., primarily methane). Subsequent reactions, such as the water-gas shift reaction, carbon dioxide removal, and methanation, are required. This last reaction converts carbon monoxide to methane by reaction with H_2.

$$CO + 2H_2 \rightarrow CH_4$$

Operating conditions for this exothermic reaction are 500 to 800° F and elevated pressures. The reaction is catalyzed by nickel-based catalyst.

4.2.2.2. Liquefaction of Coal

Developmental processes for the conversion of coal to petroleum liquids involve similar processing steps. Coal is usually mixed in an oil-based slurry and pumped to high pressures (e.g., 100 to 200 atm and heated to 750 to 900° F. At these conditions, hydrogen is passed through the coal slurry mixture to hydrogenate the coal slurry and convert coal to liquid. The reaction may or may not be catalyzed.

After reaction, the reactant mixture is separated into a gaseous fuel product, a synthetic crude oil, and the residue ash contained in the incoming coal. Separation of the ash from the petroleum product has proved difficult in earlier tests of coal liquefaction processes because separation conditions make conventional devices such as filters difficult to operate effectively. The most recent method for ash separation being investigated by DOE is the de-ashing technique developed by the CE Lummus Company. In this method, an antisolvent is used to effect the separation of hydrocarbon product from a steam rich in ash content. The major advantage of this process is that it requires no mechanical parts to effect separation.

Another method of coal liquefaction involves the gasification of coal first. The synthesis gas (mixture of $CO + H_2$) can then be reacted to produce a synthetic gasoline. This is currently practical by the SASOL Corporation of South Africa. An alternative to this method is a conversion of the synthesis gas to methanol as earlier described. The methanol may then be used directly as a replacement for gasoline in cars of the future, or it may be converted to gasoline via a catalytic process being developed by the Mobil Corporation.

4.2.3. Oil Shale Liquefaction

As in conversion of coal to petroleum liquids, hydrogen can be used to upgrade the liquefied product by suitably adjusting the carbon to hydrogen weight ratio. Reaction conditions would be similar to those of coal liquefaction.

4.2.4. Tar Sands Upgrading

In certain parts of the world (Canada, Venezuela) large reserves of tar sands exist.

These reserves are various mixtures of sand, clay, minerals, water, and crude bitumen, a substance similar to asphalt. To obtain synthetic oil from tar sands by modern techniques, the sands must first be mined. The bitumen is then separated from the sand before it undergoes a series of processing steps that convert it to a synthetic oil. A key processing step will be one similar to hydrocracking whereby the bitumen is subjected to high pressures and temperatures and reacted with hydrogen. Under these conditions the bitumen is cracked to yield a lighter product that can be sold as synthetic oil. Depending on the nature of the tar sands, a catalyst may not be required since the minerals content of the sands may have sufficient activity to catalyze the reaction.

4.2.5. Ethylene Glycol

Researchers, in in attempt to extend the usage of coal resources, are currently examining the potential for coal-derived chemicals. One petrochemical product currently derived from petroleum resources which may some day be based upon hydrogen is ethylene glycol. Ethylene glycol is used in the production of man-made polyester fibers; it is also used almost exclusively in the production of automotive antifreeze products. In current technology, ethylene glycol is produced by reacting oxygen with ethylene to produce ethylene oxide. The oxide is hydrolysized to form the glycol ($C_2H_6O_2$):

$$C_2H_4 + \tfrac{1}{2}O_2 \rightarrow C_2H_4O$$

$$C_2H_4O + H_2O \rightarrow C_2H_6O_2$$

In the future, coal may be gasified to form a synthesis gas rich in CO. After undergoing a shift reaction to alter the H_2 to CO ratio, the synthesis gas would be subjected to specific reaction conditions (e.g., pressures in excess of 500 atm) so that the following reaction occurs:

$$3H_2 + 2CO \rightarrow C_2H_6O_2$$

At this time, the reaction is still in the laboratory testing stage.

4.2.6. Cryogenic Applications[3]

Liquid hydrogen has the second lowest boiling point of any liquid. Until recently no materials had been found that were superconducting above 20.4 K, so there was no possibility of using liquid hydrogen to cool superconducting equipment. Recently, however, it has been found that Nb_3Ge superconducts at temperatures up to about 23 K. Thus, there is widespread hope that other materials will be found with even higher superconducting transition temperatures, thereby opening the way to use liquid hydrogen as a coolant instead of the usual liquid helium.

This recent discovery is very important for several reasons. First, it takes more energy to liquefy helium than hydrogen. (The theoretical minimum for helium is about twice that of hydrogen). Second, helium is a rare element mainly found in a dilute (2%) association with natural gas from a few fields. If additional and useful new materials are found that will superconduct in liquid hydrogen, important new uses for the hydrogen will be established. Important synergisms with other aspects of the hydrogen economy would be inevitable.

4.3 HYDROGEN CONSUMPTION

Table 3 is a tabulation of typical hydrogen requirement rates for the processes previously described.

REFERENCES

1. **Kelley, J. H. and Tasimann, E. A.,** Hydrogen Tomorrow...Demands and Technology Requirements, report of the Hydrogen Energy Systems Technology (HEST) Study for NASA, Jet Propulsion Laboratory Report JPL 5040-1, December 1975.
2. **Escher, W. J. D.,** Hydrogen Utilization in the Industrial Sector, paper presented at Annual Canadian Society of Chemical Engineers Annual Conference, Toronto, October 1976.
3. The Hydrogen Economy, paper prepared for Research Applied to National Needs, National Science Foundation, Grant ERP73-02706, Stanford Research Institute, Palo Alto, Calif., February 1976.

Chapter 5

Safety

Chapter 5

SAFETY*

F. J. Edeskuty

TABLE OF CONTENTS

5.1. INTRODUCTION

Although hydrogen had been discovered by Paracelsus in the early 16th century, it was not until the mid- to late-18th century that further investigation led to a better understanding of its properties. Cavendish and Priestly at first thought of hydrogen (or inflammable gas, as it was then known) as pure phlogiston. After further experiments and the passing of the phlogiston theory, Lavoisier eventually suggested the name of hydrogen.

About this same time (the 1780s) interest in hydrogen began to spread, and with this a need became obvious for a more thorough understanding of safety in handling hydrogen. Pilatre de Rozier chose to demonstrate the flammable properties of hydrogen by breathing it and then setting fire to the gas as he exhaled it through a glass tube. On one occasion, he is reported to have repeated this experiment with a mixture of hydrogen and air. The explosion which followed left de Rozier thinking that he had blown out all of his teeth.[1]

Although surviving this incident, de Rozier was by no means through with hydrogen. In 1784, he designed and had built a balloon which was to combine the advantages of both the hot air and hydrogen balloons. The danger of an open fire underneath a hot-air balloon with a hydrogen balloon above was pointed out, but de Rozier apparently felt that any leaking hydrogen would rise away from the flame below. In 1785, de Rozier was killed when the hydrogen balloon caught fire with a flash of flame and the balloon crashed to the earth. Thus died the first aeronaut.[2]

*   Work on this chapter was performed under the auspices of the United States Energy Research and Development Administration (USERDA) at Los Alamos Scientific Laboratory, University of California, Los Alamos, N.M.

TABLE 1

Properties of Hydrogen Related to Safety

Combustion range (vol % in air)	4—75
Combustion range (vol % in O_2)	4—94
Detonation range (vol % in air)	18.3—59
Detonation range (vol % in O_2)	15—90
Ignition temperature in air (K)	847
Minimum ignition energy in air, (mJ)	0.02
Minimum absolute pressure for ignition in air	
$\quad$ N/m^2	$\sim$67[a]
$\quad$ Torr	$\sim$50[a]
Flame temperature (K)	2323
Flame velocity (cm/sec)	270
Flame emissivity	0.10
Quenching distance (at 1 atm) (cm)	0.06
Heat of combustion (kJ/g mol)	242
Diffusion coefficient in air at 273 K (cm^2/sec)	0.63
Electrical resistivity of LH$_2$ (Ω-cm)	$\sim$4.6 × 10^{19}
Volume ratio (gas at 1 atm and 300 K to LH$_2$)	860
Pressure to maintain LH$_2$ density in gas at 300 K (atm)	$\sim$2000
Normal boiling temperature of LH$_2$ (K)	20.3
Volume latent heat of vaporization (J/ℓ)	31,700

[a] See text.

A more recent and better known tragedy is that of the dirigible Hindenburg. The burning of this lighter-than-air craft upon its landing at Lakehurst, N.J. in May 1937 was photographed, and the spectacular motion pictures which were widely shown in motion picture theater newsreels helped to arouse a general consciousness of the hazards of handling hydrogen. However, it is felt by many that this consciousness has been overemphasized to the point where it is sometimes referred to as the "Hindenburg syndrome", and this inordinate fear has delayed or prevented developments where the use of hydrogen might have been safe and advantageous.

Thus, in approximately a century and a half of using hydrogen, we have evidences both of disregard for safety and an overly defensive attitude. It has been said that *anything* that is combustible can be made to explode, and any gas that is not oxygen can be made to asphyxiate.[3] Therefore, to handle hydrogen safely, one must be aware of the potential hazards of hydrogen as well as have a thorough knowledge of its properties. Only then can an intelligent assessment be made concerning the risk of its use.

5.2. PROPERTIES OF HYDROGEN THAT AFFECT ITS SAFETY OF HANDLING

Properties of hydrogen that are of concern to safety are listed in Table 1. Some of these properties indicate a greater hazard potential for hydrogen when compared to other combustible fluids such as gasoline or methane (natural gas). Examples of such properties are the high flame velocity, low ignition energy, and small quenching distance. On the other hand, the high ignition temperature gives hydrogen a favorable comparison. In fact, a lighted cigarette or cigar will not ignite hydrogen in air. However, a ban on smoking in areas where hydrogen can accumulate is still necessary because a match or cigarette lighter will ignite a hydrogen-air mixture.

Properties which are frequently mentioned as making hydrogen especially hazardous are the wide combustibility and detonability ranges. Usually the safety of a given situation will require the maintenance of the atmosphere below the lower explosive limit. Since the lower explosive limit of hydrogen is similar to that of methane (5%) and even higher than that of gasoline vapor, hydrogen is not necessarily more dangerous to handle and, in general, the wide range of combustibility is not necessarily detrimental to safety in the handling of hydrogen.

The high diffusivity of hydrogen in air can make hydrogen either more or less safe than slower diffusing combustible gases. In confined spaces, hydrogen will mix faster and reach all parts of the space more rapidly. However, it will also dissipate faster with the result that, for a given hydrogen release, at any one time only a small portion of the hydrogen may exist in concentrations within the combustibility limits. It is for this reason that the total energy release from a hydrogen explosion is usually estimated as only a few percent of the total energy available.[4] The low flame emissivity is also ambiguous in its safety considerations. A hydrogen flame will radiate less energy, thus reducing the likelihood of causing burns to personnel at a given distance from the flame. On the other hand, the low emissivity makes the flame invisible (in the absence of impurities being heated in the flame) and, thus, presents the additional hazard of someone inadvertently entering a hydrogen flame.

The last five properties listed in Table 1 are of concern in handling liquid hydrogen (LH_2). The extremely high electrical resistivity reduces the likelihood of electric charge accumulation in flowing LH_2, although the flow of two-phase hydrogen can allow the accumulations of some charge which can persist.[5] As with any cryogen, the large volume ratio between liquid cryogen and room temperature gas (or conversely, large pressure necessary to maintain liquid density at ambient temperature) gives rise to a mechanism for a large pressure build-up in closed containers. The low normal boiling point (20.3 K for equilibrium LH_2, see Volume III, page 20) implies the existence of a large thermal driving force attempting to force heat from the surroundings into the LH_2. Since there is no perfect insulation, this heat input can only be minimized (or avoided by an additional refrigeration process). In addition, the low volume latent heat of vaporization indicates that larger quantities of stored liquid will be evaporated per unit of energy input.

Finally, it should be mentioned that the flammability limits given are for a system pressure of 1 atm. For hydrogen, these limits are relatively independent of pressure as pressure is decreased until, at about 0.3 atm, the combustibility range begins to narrow and finally shows no combustibility at absolute pressures below 50 torr. However, since the quenching distance increases as total system pressure decreases, this minimum combustion pressure has been questioned since this apparent limit may be a function of equipment size.[6]

5.3. HAZARDS

The above properties form the basis for the potential hazards that can be encountered in the handling of hydrogen. The most widely recognized hazard is that of unwanted combustion. For combustion to occur, it is necessary to have a fuel, an oxidizer, and an ignition source in sufficient contact. For hydrogen handling, control measures attempt to eliminate two of these three components at all times except in the case of intentional flaring for hydrogen disposal.

This means that measures must be taken to avoid hydrogen-air mixtures both within the handling equipment as well as external to this equipment. In addition, ignition sources must be avoided, which is difficult in hydrogen systems because the ignition energy is very low. Also, many experimental uses of hydrogen have required the use of auxiliary electrical and electronic equipment in which it is very difficult to eliminate all of the ignition sources. Frequently static electricity, which can even arise from solid impurity particles traveling through the piping system, can cause ignition. In the case of hydrogen being exhausted to the atmosphere, ignition control is not completely possible since electrical storms can cause ignition.

Combustion of hydrogen within closed spaces can cause serious pressure escalations. In the case of an unconfined ignition of an H_2-air mixture, a simple combustion or deflagration usually occurs resulting in overpressure up to 7 kPa. However, a confined ignition of hydrogen will cause much higher pressures. A simple, confined deflagration can raise pressure by a factor of eight while a detonation, with its supersonic reaction speeds, will give rise to much larger and faster pressure rises to pressure ratios on the order of 15:1. The possibility of a detonation depends upon the hydrogen concentration (see Table 1) as well as the strength of the ignition source. In the case of a confined deflagration transiting to a detonation, there exists the possibility of the two processes combining in such a way as to result in the product of these pressure rises with a final pressure ratio of 120:1.[7] In addition to the damage from a combustion caused by such overpressures, hazards also arise from the flame, either from direct contact or from thermal radiation, smoke, or shrapnel.

Another potential hazard that can be caused by gaseous hydrogen at ambient temperatures is the weakening of the structural members of equipment by the physical embrittlement of these solid structures by the hydrogen. This phenomenon, which is essentially a room-temperature phenomenon, is called hydrogen embrittlement and is discussed in Volume II, Chapter 4.

Additional hazards can arise in the handling of LH_2. As with any cryogen, there is an unavoidable heat leak into the storage apparatus which requires that either additional refrigeration must be supplied to match this heat input or the boil-off gas must be allowed to vent (usually to the atmosphere). If the resulting gas is not vented, the pressure will rise until either a relief valve opens, an equilibrium pressure is reached, or the vessel ruptures. The pressure rise rate of nonvented LH_2 dewars has been studied.[8] For cases where thermal equilibrium is established within the dewar, the rate of pressure rise can be estimated by calculation using the first law of thermodynamics. However, LH_2 dewars will usually stratify thermally, thus giving rise to pressure rise rates up to an order of magnitude higher than those calculated with the assumptions of complete thermal mixing.

Some smaller dewars have thermal radiation shields that are cooled by the venting vapors, thus gaining a lower boil-off loss by utilizing some of the refrigeration available in the cold venting vapor. For such a dewar, the stopping of venting for any reason will result in an increased heat leak (by up to, perhaps, a factor of 10) and consequent faster pressure rise.

LH_2 is sufficiently cold that, if any air is allowed to enter the storage vessel or its vent system, the air (in addition to any moisture it contains) will condense to a solid and exert essentially zero vapor pressure. This is a potential hazard for several reasons. First, it forms a combustible mixture. Also, the air so condensed is very difficult to detect so that a considerable amount of accumulation can result. In addition, a large fraction of this air can condense within the storage vessel vent tube where the temperature is at the freezing point of air. A sufficient accumulation at this point can plug the vent tube, thus starting a pressure build-up which eventually can result in a catastrophic failure of the storage vessel. This type of accident has occurred.

The low temperatures encountered in cryogenic hydrogen systems (typically down to 20 K [−425°F]) can cause hazards in other equipment and in structural members. One must consider potential hazards arising from changes in physical properties. In addition to the obvious necessity of allowing for thermal contraction, there is also the possibility of embrittlement due to the low temperature. While 9% nickel is satisfactory for use with liquid nitrogen and liquid oxygen, it is not satisfactory for LH_2 service.

Air condensation can also be dangerous when it occurs externally to the LH_2 system. Cold hydrogen gas or LH_2 being vented through uninsulated pipes will usually condense liquid air on the outside of the pipes. If this air is allowed to drip onto a combustible surface, such as asphalt, an explosive mixture can result. For this reason, LH_2 unloading or transfer areas are usually paved with concrete instead of asphalt. The dripping liquid air can also be hazardous if allowed to fall onto other equipment that can be disabled by being cooled too much, too unequally, or too rapidly. This phenomenon has cracked deck plates on LH_2 transport barges and even the outer shell (carbon steel vacuum jacket) of a vessel from which cold hydrogen was being discharged.

It should be pointed out that, even without cooling from LH_2 being vented, there frequently exist large temperature gradients within equipment handling LH_2, and care must be taken to avoid excessive thermal stresses as well as the inadvertent cooling of any equipment which could be damaged by cooling. This could arise either from cold embrittlement (a large loss of ductility as temperature is lowered) or thermal contraction in excess of that allowed for in the design of the equipment.

The piping system and storage vessels in LH_2 systems are obvious places where temperature gradients must be considered in both the design and the planning of operating procedures. Temperature gradients will always exist in the cool-down to operating temperature, and frequently the residual temperature gradient remaining after reaching thermal equilibrium can also give rise to prohibitive stresses.

An additional possible hazard is that of asphyxiation in a H_2 atmosphere. However, the purging requirements and high diffusivity of H_2 tend to minimize the likelihood of such a hazard arising. With handling of LH_2, there is the additional hazard of frost bite to personnel, especially if cold, uninsulated vents or transfer pipes are left exposed.

5.4. AVOIDANCE OF HAZARDS

The above hazards and how to avoid them must be considered in system design and construction as well as in planning the operation of the system. The equipment must be built so that it allows safe operation.

To avoid unwanted hydrogen-air mixtures, provision must be made to allow the purging of air from the equipment before admitting the hydrogen. After operation, it is also advisable in most cases to purge the remaining hydrogen from the system. For equipment with a low length to diameter ratio, such as a spherical dewar, it frequently is best to evacuate the vessel to a residual pressure of no more than a few torr and then backfill with either hydrogen or an inert gas such as nitrogen or helium. If directly backfilling with hydrogen, the evacuation should be to a pressure low enough that there is no chance of creating an explosive mixture at any time during the purging process. One can also pressurize the container to some elevated pressure with an inert gas and then vent the resulting mixture, repeating this procedure until a sufficient purge has been obtained. Either of these procedures can take a number of cycles before a satisfactory purge is obtained.

For equipment with a long length to diameter ratio, it is frequently better to use a flow-through purge. An example of this type of equipment is a transfer line or pipe. Care must be taken to avoid dead ends in the piping system, and where they must exist, some of the purge gas must flow out an additional vent at the end of the side branch.

For any purging operation, one should not rely upon calculation of the degree of purge obtained. Rather, samples should be taken from several points of the system and analyzed for residual traces of air. A gas chromatograph is a convenient and accurate analysis tool and can usually detect oxygen and nitrogen down to a few parts per million. However, to obtain this degree of accuracy, it is necessary to be very careful in the sampling procedures so that residual or adsorbed air in the previously open portions of the sampling system does not result in abnormally high air concentrations in the sample. If a "blow-through" type of sampling is to be done, the purge gas should be allowed to vent through the sample bottle for about 10 min before closing it off.

It is the preferred procedure to contain all hydrogen within the using system. If small amounts of leakage do occur, adequate ventilation can assure that no explosive mixtures will occur. However, in some cases where very large amounts of hydrogen are involved and the possibility of large leaks exist, it might not be possible to provide adequate ventilation. This was the case where LH_2 flowed through the flow control room in the nuclear rocket development program.[9] In this case, it was decided that the entire room should be inerted by blowing in nitrogen gas until the concentration of oxygen was below 4%. The oxygen concentration was then monitored, and additional nitrogen was introduced whenever the oxygen concentration reached 4%. In order for any personnel to enter the room with this procedure, it was necessary that either deinerting take place or the entering personnel wear self-contained breathing apparatus. Measurements of the quantity of nitrogen gas required to effect the inerting confirmed the assumption of complete mixing of the input gas, and thus, 1.7 room volumes of gaseous nitrogen were required to perform the initial inerting.[10]

With the use of hydrogen systems, it is always necessary to control ignition sources. This begins with the prohibition of smoking, open flames, etc. within specified distances. Elimination of ignition sources from other equipment is more more difficult to control. Explosion-proof equipment is equipment that will contain an ignition within the case. However, due to the small quenching distance for a hydrogen flame (see Table 1), the manufacture of such equipment is difficult. Therefore, if equipment which could become an ignition source must be located within an area where combustible mixtures of hydrogen and air could occur, this equipment is frequently contained within an enclosure which is purged with nitrogen gas in a manner analogous to the room inerting mentioned above. In cases where it is necessary to dispose of hydrogen continuously (LH_2 storage is such an example), ignition can occur from natural causes. A direct lightning strike is not necessary since the corona discharge accompanying lightning striking within hundreds of feet can ignite the venting gas. LH_2 storage systems should be equipped with vent systems that allow for such a fire with no damage resulting. Check valves in the vent line will prevent backflow of air, even when venting ceases. The vent stack can be equipped with a helium supply. In the event of a stack fire, a slow helium flow can be started, next the hydrogen can be valved off, and the helium allowed to continue to flow until the top of the vent system has cooled sufficiently that no reignition will occur. This will take about 5 min. Then the hydrogen venting can be restarted, and finally the helium flow can be stopped.

In the case of large quantities being vented, it is not always best to extinguish a fire. If the venting cannot be stopped, the possibility of hydrogen accumulation may present a greater hazard than the existing fire.

For routine venting of LH_2 storage dewars, the flows are usually small enough that the venting can be directly to the atmosphere with no provision for flaring the vented hydrogen gas. Under normal circumstances, this venting will only amount to about 1 g/sec. However, for the venting of railroad LH_2 storage tank cars with similar boil-off losses, a venting system was devised that mixed air with the exhausted hydrogen in sufficient quantities that the vented mixture was always below the combustible limit in hydrogen concentration.

Except in unusual circumstances, LH_2 should be stored at a pressure above that of the surrounding atmosphere so that any leakage will result in hydrogen escaping rather than air entering the storage dewar. For low boil-off dewars, the vent system should make provision for preventing backflow of air. Also, the velocity within the vent system should be maintained high enough so that appreciable quantities of air cannot enter the dewar by the slow back-diffusion of air through the stagnant boundary layer of hydrogen adjacent to the vent tube wall.[11]

If large quantities of hydrogen must be vented in a short time (several hundred grams per second or higher), the usual procedure is to flare. One method of flaring makes use of a flare stack incorporating a molecular seal, which is shown schematically in Figure 1. The arrangement shown allows a very small flow to exclude the atmosphere from the interior of the stack. Even under windy conditions, a small flow of gas is effective in keeping combustible mixtures out of the stack. Vent stack stability has been studied, thus making it possible to predict flow rates for stable operation without flame blowoff or burnout.[12] Another method of disposing of large quantities of hydrogen is by use of a burn pond. With this method, the hydrogen is released by being allowed to bubble up through a pond of water and burn at the surface of the water.[13]

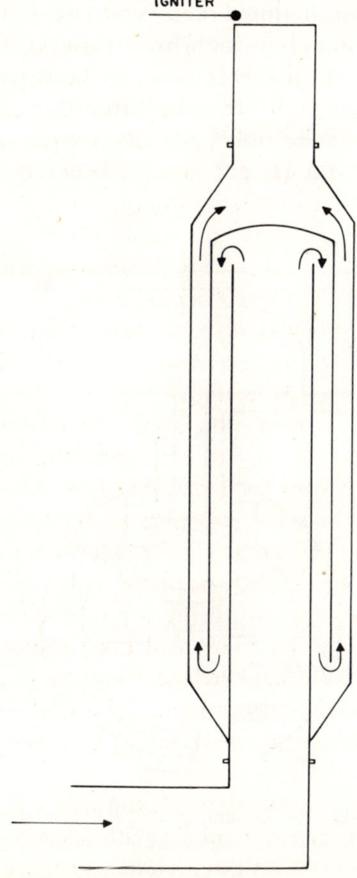

FIGURE 1.   Hydrogen flare stack.

Venting of cold (or liquid) hydrogen is usually accompanied by a cloud of condensed water vapor. The limit of the range of combustible mixture has been shown to exceed the cloud volume;[12] thus, it is not safe to assume that the cloud filled area is the only area where caution must be taken.

Before leaving the subject of hydrogen disposal, it should be noted that, as a rule, it should be attempted to make each storage dewar for LH_2 independent of the rest of the system. This means that the normal boil-off losses should be vented from a local vent stack completely separate from the rest of the system. In this manner, there is no interaction that could cause a problem in the dewar due to any problem in some other part of the system.

In any system containing either LH_2 or cold hydrogen gas, it is necessary to provide for pressure relief to prevent any unwanted pressure rise when the system finally warms up to ambient temperature. Large LH_2 storage dewars are usually equipped with a venting system similar to that shown in Figure 2. Vent systems for smaller dewars may look somewhat different but should incorporate the same features, including a minimum of two entries to the dewar interior. The system shown in Figure 2 provides for normal pressure relief at a preset pressure level backed up by a safety pressure relief valve and finally by a rupture disc. The safety relief valve should be checked periodically, and the rupture disc should be replaced at preset intervals, since it can either suffer from work-hardening (which could raise its relieving pressure) or from fatigue (which could lower its relieving pressure). If dewar pressurization gas must enter the dewar, it frequently is necessary to provide within the dewar a diffuser inlet that prevents the incoming gas from impinging directly upon the liquid surface.

The storage dewar is not the only place where it is necessary to provide for pressure relief. Figure 3 shows schematically three types of spaces that require pressure relief. Within a storage dewar, the necessity for provision of a relief valve is obvious. However, the same sort of requirement holds for any section of transfer line or any container where either LH_2 or cold H_2 gas can be isolated between two valves, either

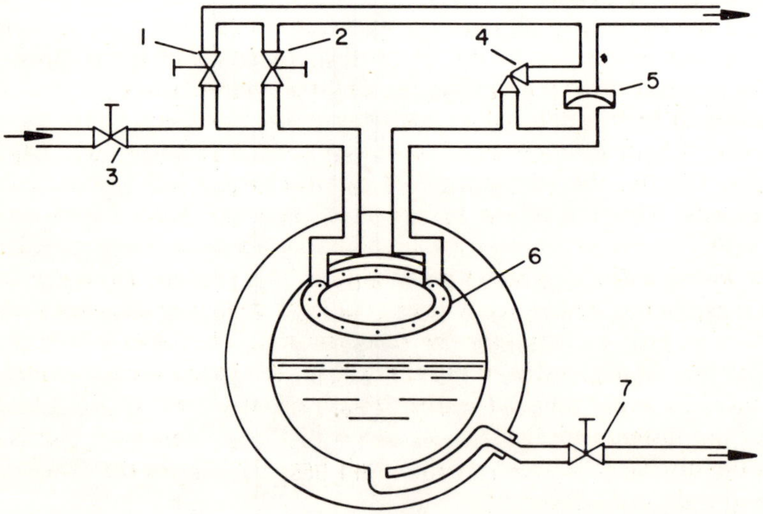

FIGURE 2. LH_2 storage dewar vent system. (1) Normal boil-off vent valve. (2) High rate vent valve. (3) Valve to admit dewar pressurization gas. (4) Safety pressure relief valve. (5) Rupture disc. (6) Dewar pressurization and vent ring. (7) Liquid hydrogen discharge valve.

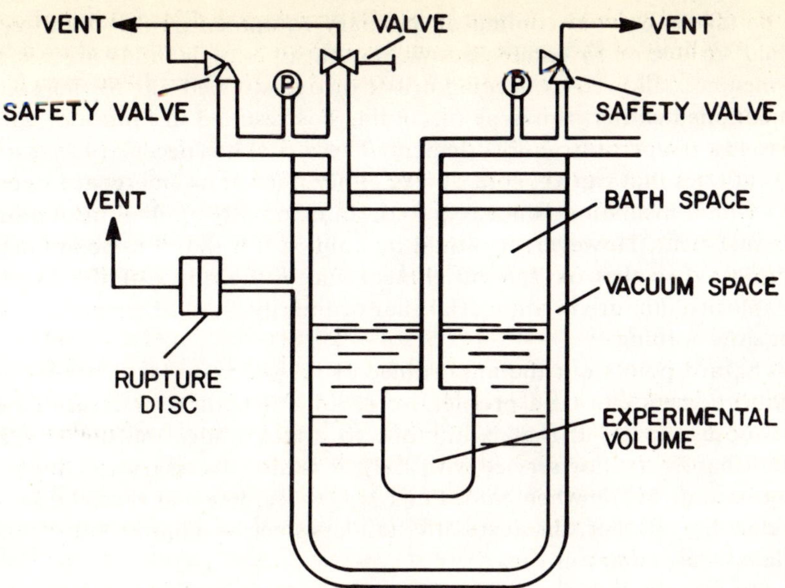

FIGURE 3.   Volumes that require pressure relief.

intentionally or inadvertantly. A second type of volume which requires similar protection is an internal experimental gas volume which relies on the external LH_2 to keep it cold. Here, loss of refrigeration could cause unwanted pressure build-up. Finally, it should be remembered that a third volume — the vacuum insulation space which is usually used to surround cryogenic systems — is another space that needs the protection of a pressure relief valve. The cold inner surface of the vacuum space is an excellent cryopump for any air that might leak in. At 20 K the vapor pressure of N_2 is approximately 10^{-11} torr,[14] and therefore, fairly large quantities of air that could have leaked in over a long period of time would not be detected. Upon removal of the cryogen from the system, its subsequent warm-up could then cause a rapid pressure rise in the vacuum space due to the consequent warming of the trapped air. Thus, provision for pressure relief is necessary for all three types of space.

The handling of hydrogen gas does not present any low temperature problems, as a rule. However, if hydrogen gas which is stored at high pressure (say, 100 atm) is allowed to vent rapidly, the temperature of gas discharged will become colder as the venting proceeds. This might not be expected since the Joule-Thomson expansion across the vent valve is at a temperature above the inversion temperature. Thus, the gas actually warms a few degrees as it expands by this process. However, the gas still in the tank is expanding adiabatically (since the thick tank wall does not permit a rapid enough influx of heat to maintain the temperature). The latter effect can cool the remaining gas tens of degrees, with the result that the exhaust stream constantly drops in temperature. To avoid the cold embrittlement of tank farm vessels which were not intended for use in temperatures below, say, $-30°C$, stainless steel inserts have been installed in the discharge tube where the rapid heat transfer to the flowing gas could cool the vessel walls excessively.

Use of LH_2 brings additional possibilities for creating cold surfaces where they are neither expected nor wanted. Such cold metal surfaces can present greater burn hazards to personnel than the cryogen itself does. At hydrogen temperature, these cold surfaces can condense liquid air. The air so condensed is enriched in oxygen and, therefore, gives rise to even greater hazard if allowed to drip onto combustible objects. Care must also be taken to avoid cooling of auxiliary equipment, either by direct contact with the cold surface or by condensed air, so that operating fluids are not frozen or structural members allowed to become brittle or overstressed due to thermal contraction. As mentioned earlier, this type of cooling has resulted in failure of carbon steel plates due to low temperature embrittlement. Therefore, it is necessary to insulate thermally the vent lines that can get cold during either normal or emergency operation. In this case, vacuum insulation is not required; an externally applied insulation, such as a foam, is sufficient. However, it should be pointed out that a noncombustible foam is to be preferred so that oxygen enrichment within the pores of the foam does not form an explosive mixture in a material that ordinarily would be considered noncombustible or slow burning.[15]

The last hazard points out the importance of proper material selection. Where hydrogen embrittlement can be a problem, e.g., in a structural material, care must be taken to choose a material that is immune to attack under the use conditions (see Volume II, Chapter 4). For service with LH_2, it is also necessary to choose materials that do not become brittle when cold. For LH_2 service, 9% nickel steel does not retain sufficient ductility. Rather, the austenitic stainless steels — copper, brass, and the aluminum alloys — are commonly used.

Needless to say, all equipment should be pretested to establish the ability to function properly, this testing to be done with an inert or otherwise safe testing gas. In the design of equipment, it is usually best to use the more conservative ambient temperature values for yield and tensile strengths. Not only will this result in a more conservative design, but also it will allow for the fact that it is frequently necessary to perform pressure tests at ambient temperature. Also, it must be kept in mind that, within cryogenic equipment, there frequently exist large temperature gradients which not only make more appropriate the use of the lower strengths but also can add thermal stresses to the operating stresses.

For systems which will experience large temperature gradients (e.g., LH_2 use), the large thermal contractions that can be experienced necessitate a careful analysis of the thermal stresses that can arise. The thermal contraction from room temperature to cryogenic temperatures amounts to approximately 0.3% for stainless steels and copper, 0.4% for aluminum, and well over 1% for some plastics.[16] Only Invar® has a small enough thermal contraction that this mechanism for building up stress can sometimes be ignored. Table 2 presents some values of this integrated thermal contraction between ambient and cryogenic temperatures (20 K). For long piping systems either there must be sufficient flexibility by means of U bends and elbows, or the pipe must be provided with expansion bellows. While the method of computing these stresses is straightforward, for large systems it is usually tedious and requires the use of computers to allow the inclusion of all the necessary constraints and supports.[17]

For piping systems at equilibrium conditions, the stresses can be computed with the knowledge of the geometry of the system, the material properties, and the operating temperature. However, cryogenic systems must be cooled to operating temperature and, during the cooling process, the mechanism exists for powerful transient cooling stresses. For high-pressure systems, the thickness of flanges can be great. For example, an 8-in., schedule 40 flange has an outside diameter of 20 in. The thickness of this flange can allow a cooldown procedure such that the inner bore of the flange will be at 20 K while the outside surface can still be at ambient temperature. This will usually overstress the flange material. This problem has been analyzed, and maximum flow rates for cooling thick wall pipes are given in Figure 4.[18] The curves in Figure 4 are calculated assuming that the maximum thermal stress will be no larger than the yield stress of the material at room temperature.

TABLE 2

Values of Integrated Thermal Contraction of Some Selected Materials Between Ambient and LH₂ Temperature (293 K to 20 K)

Material	$\dfrac{L_{293}-L_{20}}{L_{293}}$
Stainless steel (AISI304)	0.00296
Copper	0.00326
Aluminum	0.00415
Invar®	0.00052
Nylon[a]	0.01379
Teflon®[a]	0.02110
Pyrex®	0.00056

[a] Note that the thermal contraction of plastics can be greatly modified by filling with glass, graphite, etc. Such composites can be made to match most metals very closely in thermal contraction.

From Corruccini, R. J. and Gniewek, J J., *Nat. Bur. Stand. U.S. Monogr.*, No. 29, 1961.

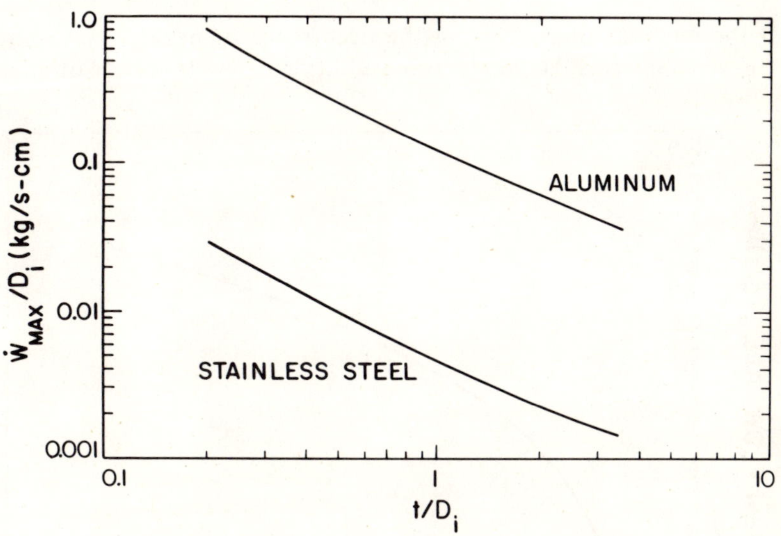

FIGURE 4. Maximum safe flow rates of LH₂, W_{MAX}, for transfer line cooldown. W_{MAX} D_i = inside diameter and t = wall thickness (cm).

In the cooldown of cryogenic systems, trouble can also be encountered from too slow a cooldown rate. In the cooldown with LH_2, some two-phase flow must be encountered since the cooling fluid is introduced as a low-pressure liquid (below the critical pressure), and this fluid is eventually vented from the system as a room-temperature gas. In the portion of the pipeline with two-phase flow, there exists the possibility of several types of flow. For low flows, the two phase can stratify with the liquid flowing on the bottom of the pipe and the gas on top.[19] The available heat of vaporization (plus better heat transfer to the liquid) provides a mechanism for the bottom of the pipe to cool faster than the top.

The result of these two restrictions is that, upon cooling a system to operating temperature with a cryogenic fluid, the operation must be such as to stay within a narrow operating range between the maximum shown in Figure 4 and the minimum shown in Figure 5,[20] where the flow indicated is sufficient to maintain nonstratified flow until 95% of the liquid has been vaporized. In some cases, there may be no acceptable LH_2 cooldown flow, and other methods, such as cold gas precooling, must be used.

For the safe operation of any hydrogen system, it is necessary to have the proper instrumentation. This instrumentation should not be an afterthought, but rather should be considered early in the conceptual design of the system. In this way, there will have been provided the necessary sampling ports as well as the hydrogen-monitoring equipment and gas analyzers. Frequently, it is much more expensive and sometimes impossible to add desired temperature measurement points or sample ports once the equipment is built. It should be pointed out that hydrogen is colorless, tasteless, and odorless. Furthermore, LH_2 is a very volatile liquid, and to cut the heat leak into the system, every effort is made to cut all radiation to it, even visible light radiation. For this reason, in a well-designed and operated system, LH_2 is seldom, if ever, seen. This characteristic of hydrogen makes the operator completely dependent upon the instrumentation for the safe, reliable, and efficient operation of the system.

5.5. ADDITIONAL DESIGN AND OPERATIONAL DETAILS

Some of the methods of preventing the occurrence of hazards have already been mentioned in general terms. However, some additional details may be of interest.

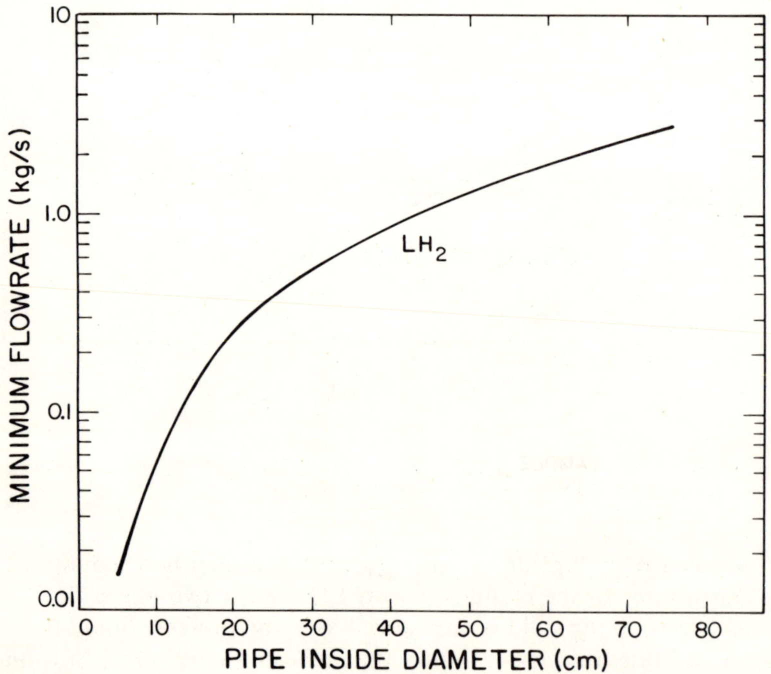

FIGURE 5.   Cooldown minimum flow rate of LH_2 to avoid stratified flow until 95% of the liquid is vaporized.

In some cases, hydrogen systems must be safely isolated from other systems. For gaseous hydrogen, the ring and blank (or "figure 8") flange is satisfactory (see Figure 6a). By making use of this double insert with one side open and one side a solid disc, the condition of the line is obvious at a glance. However, there is the disadvantage that a joint must be opened to swing the ring from one position to the other. Another solution is the "double block and bleed" in which two valves are placed in series, and the volume between these valves is vented to the atmosphere (see Figure 6b). For the isolation of the liquid discharge line of an LH_2 storage dewar, neither of the above methods is satisfactory.

For this purpose a so called "helium block" has been used (see Figure 6c). Here, two liquid shut-off valves are placed in series. After an LH_2 discharge is completed, the line is purged with helium gas, and the space between the two valves is charged with helium gas at a slightly higher pressure than the LH_2 storage dewar. If a double-block-and-bleed method were attempted, a small amount of leakage out of the dewar would first destroy any gas block (or trap) which was included in the system to minimize heat leak from the shut-off valve into the dewar. Then the cooling of the valve would increase the leakage as the valve becomes progressively colder. With the helium

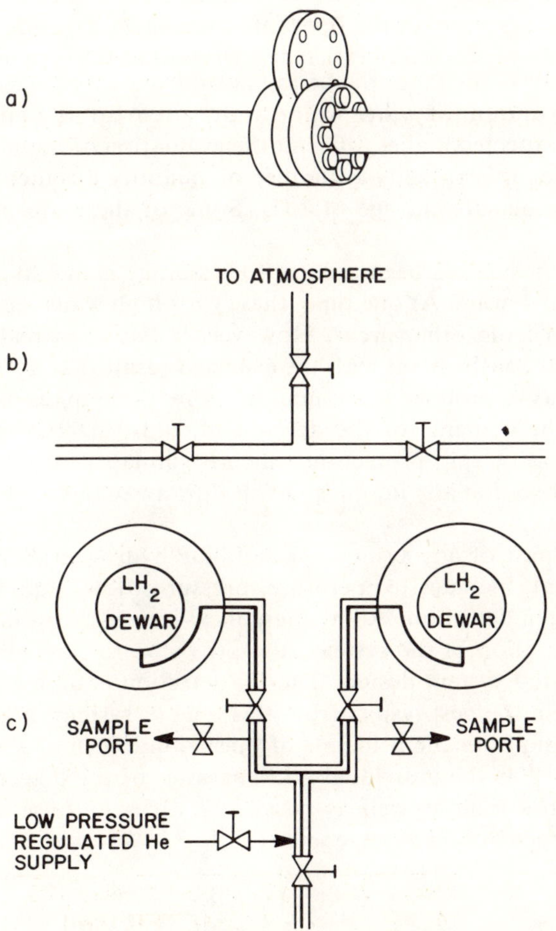

FIGURE 6. Hydrogen shut-off systems. (a) Ring and blank. (b) Double block and bleed. (c) Helium block.

block, any small amount of leakage will be only of helium gas and into the dewar. It should be mentioned in passing that a hydrogen system should be made safe under the assumption that any valve in the system can leak at any time.

The low vapor pressure of nitrogen at 20 K which was mentioned earlier also makes it very difficult to detect the presence of solid air within an LH_2 storage volume. For this reason, in some cases it has been advisable to periodically warm a storage dewar to approximately 100 K and then analyze the contents to look for O_2 and, thus, determine whether any significant quantities of dangerous contaminants are accumulating.

Another subject which is of interest in the large-scale use of LH_2 is the safe distance required for storage, either from LH_2 dewar to inhabited buildings or between LH_2 dewars. Considerations which affect such a quantity-distance relationship are possible fireball size, flame radiation, shrapnel damage, and overpressure. The size of fireball has been given as:[7]

$$D = 7.93\,W^{1/3}$$

where D = diameter (m) and W = mass of hydrogen consumed (kg). Duration of burning (in seconds) has been given as:[7]

$$t = 0.47\,W^{1/3}$$

However, the intensity vs. distance of the radiation from a hydrogen flame is strongly dependent upon the amount of water vapor in the air. Also, the uncertainty in actual energy release to be expected makes difficult the evaluation of damage due to shrapnel and overpressure. For this reason, a number of quantity-distance recommendations have been made for the safe storage of LH_2. Some of these are presented in Figure 7.[7,21]

Another precaution that has been used in LH_2 storage is the diking of the area underneath LH_2 storage dewars. At one time, these were high walls separating the dewars from each other and from other areas. However, it was soon realized that the high walls provided partial confinement with the expected result that a spill and subsequent combustion might easily produce a detonation. Now these dikes are made only high enough to contain the contents of the dewar and, thus, are typically only 2 or 3 ft high. Another method of spill protection which is similar to diking is to channel the area under the dewar so that any liquid spill will flow away from the dewar or working area into a safer area.

A very important part of any system for handling hydrogen is proper instrumentation which, in general, includes temperature, pressure, flow, liquid level (in the case of LH_2), and hydrogen detection. All of these measurements can have a great significance in the safe operation of the system, and provision for their inclusion should be recognized in the initial system design. The H_2 detection equipment is obviously important, and to obtain the best results from this type of system, it is necessary for the operator to understand both the principle of operation (thermal conductivity or catalytic oxidation) as well as the individual idiosyncrasies of the detectors. The latter includes system response time as well as possible sources of false alarms. Also, it is usually best not to depend upon a single sensor.

5.6. SUMMARY

Constant vigilance and striving for safer, simpler operation are paramount factors in handling large quantities of hydrogen. Where these have been done, operation has

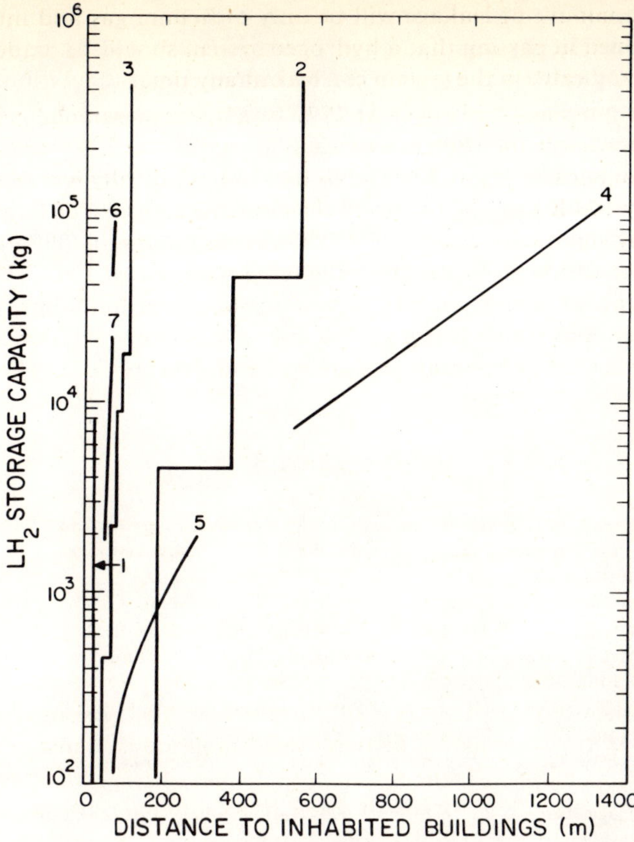

FIGURE 7.   Various quantity-distance relationships suggested for inhabited buildings to LH₂ storage. (1) From Reference 23. (2) From Reference 24. (3) From Reference 25. (4) From Reference 26. (5) From Reference 25. (6) From Reference 25. (7) From Reference 27.

been without serious incidents. In the uses which have recently been suggested for hydrogen, no new problems are indicated if known hazards are recognized and existing warnings are heeded. The most likely way that danger could be encountered is in the attempt to cut costs in very large-scale operations.

In any hydrogen operation, certain steps must be taken. Safety should be a prime consideration in all phases, including design, construction, checkout, operation, and modification. Operations should be according to preconceived plans, which include standard operating procedures as well as contingency procedures. These procedures should be written, and the documents should be reviewed regularly. The avoidance of hazards must always be considered, and the system should be operated only as planned, which includes staying within anticipated pressure, flow rate, and temperature limits.

Operator training should be an on-going practice. Operators should be encouraged to learn the reason for any restrictions or safety rules. A continuing safety program should include experiments to learn more about the safe operation of the system and to demonstrate these principles to others. There should be a continuous effort to relax unnecessary restrictions which can also be dangerous in the long run by leading to a disrespect for other safety rules that are more vital for safe operations.

Facilities which handle H₂ should be kept free from secondary hazard, such as other combustible materials, and good housekeeping is especially important. Existing codes and regulations specifically for hydrogen operations are mostly concerned with its transportation.[22] Other phases are covered by more general regulations, and a recent and excellent listing is given by Hord.[7] The references contained herein constitute good sources for additional information.

In conclusion, it should be stated that, in most uses, hydrogen is never seen. Thus, in its use there is imposed the additional burden of almost complete dependence upon instrumentation. In spite of this fact, hydrogen handling can be made safe, and it still remains an open question as to whether it is more or less safe than the more conventional fuels.

REFERENCES

1. **Friend, J. N.** *Man and the Chemical Elements*, Charles Griffin & Company Ltd., London, 1951, 32.
2. **Sweaton, W. A.,** Jean-Francois Pilatre de Rozier, the first aeronaut, *Ann. Sci.*, 11(4), 349, 1955.
3. **Reider, R.,** private communication.
4. **Edeskuty, F. J. and Reider, R.,** Liquefied hydrogen safety — a review, *ASSE J.*, 14(5), 18, 1969.
5a. **Cassutt, L., Biron, D., and Vonnegut, B.,** Electrostatic hazards associated with the transfer and storage of hydrogen, in *Advances in Cryogenic Engineering*, Vol. 7, Timmerhaus, K. D., Ed., Plenum Press, New York, 1962, 327;
5b. **Willis, W. L.,** The Electrical Conductivity of Some Cryogenic Fluids, M.S. thesis, University of New Mexico, Albuquerque, 1965.
6. **Lewis, B. and vonElbe, G.,** *Combustion, Flames and Explosions of Gases*, 2nd ed., Academic Press, New York, 1961.
7. **Hord, J.,** Is Hydrogen Safe, NBS Tech. Note 690, National Bureau of Standards, U.S. Department of Commerce, Washington, D.C., October 1976.
8. **Liebenberg, D. H. and Edeskuty, F. J.,** Pressurization analysis of a large scale liquid-hydrogen dewar, in *International Advances in Cryogenics Engineering*, Vol. 10, Timmerhaus, K. D., Ed., Plenum Press, New York, 1965, 284.
9. **Edeskuty, F. J., Henshall, J. B., and Bartlit, J. R.,** Cryogenics in the nuclear rocket program, *Cryogenics Ind. Gases*, 4(6), 36, 1969.
10. **Boyer, K., Otway, H., and Parker, R. C.,** Large volume operational room inerting, in *Advances in Cryogenics Engineering*, Vol. 10, Timmerhaus, K. D., Ed., Plenum Press, New York, 1965, 273.
11. **Buffham, B. A., Bailey, B. M., and Geist, J. M.,** Back diffusion into stored cryogenic liquids venting to the atmosphere, in *Advances in Cryogenic Engineering*, Vol. 6, Timmerhaus, K. D., Ed., Plenum Press, New York, 1961, 50.
12. **Zabetakis, M. G.,** *Safety With Cryogenic Fluids*, Plenum Press, New York, 1967.
13a. **Lapin, A.,** Hydrogen vent flare stack performance, in *Advances in Cryogenic Engineering*, Vol. 12, Timmerhaus, K. D., Ed., Plenum Press, New York, 1967, 198.
13b. **Thompson, W. R. and Boncore, C. S.,** Design and development of a test facility for the disposal of hydrogen at high flow rates, in *Advances in Cryogenic Engineering*, Vol. 12, Timmerhaus, K. D., Ed., Plenum Press, New York, 1967, 207.
14. **Honig, R. E. and Hook, H. O.,** *RCA Rev.*, 21, 360, 1960.
15. **Key, C. F. and Gayle, J. D.,** Preliminary Investigation of Fire and Explosion Hazards Associated with S-11 Insulation, NASA Tech. Memo. NASA-TM-X53144, George C. Marshall Space Flight Center, Huntsville, Ala., October 2, 1964.
16. **Corruccini, R. J. and Gniewek, J. J.,** Thermal expansion of technical solids at low temperatures, *Nat. Bur. Stand. U.S. Monogr.*, No. 29, 1961.
17. **M. W. Kellogg Company,** *Design of Piping Systems*, John Wiley & Sons, New York, 1956.
18. **Novak, J.,** Cooldown flow rate limits imposed by thermal stresses in liquid hydrogen or nitrogen pipelines, in *Advances in Cryogenic Engineering*, Vol. 15, Timmerhaus, K. D., Ed., Plenum Press, New York, 1970, 346.
19. **Baker, O.,** Design of pipe lines for simultaneous flow of oil and gas, *Oil Gas J.*, 53, 185, 1954.
20. **Liebenberg, D. H., Novak, J. K., and Edeskuty, F. J.,** AIAA Paper No. 67-475, presented at AIAA 3rd Propulsion Conf., Washington, D.C., July 1967.

21. **Edeskuty, F. J., Reider, R., and Williamson, K. D., Jr.,** Safety, in *Cryogenic Fundamentals*, Haselden, W., Ed., Academic Press, London, 1971, chap. 11.
22. **Edeskuty, F. J.,** Safety Problems and Safety Codes Concerning Liquid Hydrogen and Liquid Helium, in *Proc. 12th Int. Congr. Refrigeration,* Vol. 1, Graficas Unidas, S. A., Madrid, 1969, 283.
23. Standard for Liquefied Hydrogen Systems at Consumer Sites, Pamphlet G-5.2, Compressed Gas Association, New York 1966.
24. U.S. Department of Defense Instruction, No. 4145, 21 1964.
25. **Zabetakis, M. G., Furno, A. L., and Martindill, G. H.,** Explosion hazards of liquid hydrogen, in *Advances in Cryogenic Engineering*, Vol. 6, Timmerhaus, K. D., Ed., Plenum Press, New York, 1961, 185.
26. U.S. Army Materiel Command Safety Manual, No. 385-224, June 1964 (formerly Ordnance Safety Manual ORD M 7-224, Ordnance Corps, Department of the Army, 1951 with changes 1951, 1955, 1958.
27. Final Report on an Investigation of Hazards Associated with the Storage and Handling of Liquid Hydrogen, No. C-61092, Arthur D. Little, Inc., Cambridge, Mass., March 22, 1969.

Unit Conversions, Physical Constants, and Symbols

UNIT CONVERSIONS, PHYSICAL CONSTANTS, AND SYMBOLS

A complete description of the International System of Units is given in Page, C. H. and Vigoureux, P., National Bureau of Standards, Special Publ. 330, January 1971. A good general reference for physical constants and conversion factors is Mechtly, E. A., National Aeronautics and Space Administration, SP-7012, 1964. Selected unit conversion and physical constants are given in the following table.

CONVERSION FACTORS

Density

lb/ft^3	kg/m^3	g/cm^3	mol/cm^3	Amagat
1	16.018	0.016018	7.9458×10^{-3}	178.216
0.062428	1	0.001	4.9605×10^{-4}	11.126
62.428	1,000	1	0.49605	1.1126×10^4
125.85	2,015.9	2.0159	1	2.2428×10^4
5.6111×10^{-3}	0.089881	8.9881×10^{-5}	4.4586×10^{-5}	1

Specific Volume

ft^3/lb	m^3/kg (l/g)	cm^3/g	cm^3/mol
1	0.062428	62.428	125.85
16.018	1	1,000	2,015.9
0.016018	0.001	1	2.0159
7.9458×10^{-3}	4.9605×10^{-4}	0.49605	1

Pressure

$lb_f/in.^2$ (psi)	MPa	atm	Torr (mm Hg)	bar
1	6.8948×10^{-3}	0.068046	51.715	6.8948×10^{-2}
145.04	1	9.8692	7,500.6	10.0
14.696	0.10132	1	760.0	1.0132
0.019337	1.3332×10^{-4}	1.3158×10^{-3}	1	1.332×10^{-3}
14.504	0.1	0.98692	750.06	1

Enthalpy, Heat of Vaporization, Heat of Conversion, Specific Energies

Btu/lb	kJ/kg (J/g)	J/mol	cal/g
1	2.3244	4.6858	0.55556
0.43022	1	2.0159	0.23901
0.21341	0.49605	1	0.11856
1.8	4.1840	8.4345	1

Specific Heat, Entropy

Btu/lb-R	kJ/kg-K (J/g-K)	J/mol-K	cal/mol-K
1	4.184	8.4345	2.0159
0.23901	1	2.0159	0.48182
0.11856	0.49605	1	0.23901
0.49605	2.0755	4.184	1

Thermal Conductivity

Btu/ft-hr-R	mW/cm-K	J/s-cm-K	cal/s-cm-K
1	17.296	0.017296	0.0041338
0.057816	1	0.001	2.3901×10^{-4}
57.816	1,000	1	0.23901
241.90	4,184	4.184	1

Viscosity

lb/ft-s	kg/m-s $(N\text{-}s/m^2)$	cP $(10^{-2} g/cm\text{-}s)$	lb-s/ft^2 (slug/ft-s)
1	1.48816	1,488.16	0.031081
0.67197	1	1,000	0.020885
6.7197×10^{-4}	0.001	1	2.0885×10^{-5}
32.175	47.881	4.7881×10^4	1

Temperature

K	R	C	F
1	1.8	1	1.8

Velocity of Sound

ft/s	m/s
1	0.3048
3.2808	1

Surface Tension

lb$_f$/in.	N/m	dyne/cm
1	175.13	175.13×10^3
5.7102×10^{-3}	1	1,000
5.7102×10^{-3}	0.001	1

lb_f	=	pound force
lb	=	pound mass
g	=	gram
kg	=	kilogram mass
mol	=	2.01594 gram
ft	=	foot
m	=	meter
cm	=	centimeter
Btu	=	British thermal unit
J	=	joules = watt second
W	=	watt
cal	=	calorie
R	=	absolute temperature in degrees Rankine
F	=	temperature in degrees Fahrenheit
K	=	absolute temperature in Kelvins
C	=	temperature in degrees Centigrade
N	=	Newton
MPa	=	megapascal (pressure)
atm	=	atmosphere pressure
Torr	=	millimeter of mercury pressure
cP	=	centipoise
slug	=	32.174 lb
R	=	gas constant = 8.31434 J/mol-K, 8.31434×10^6 N-cm^3/m^2-mol-K
mol wt	=	2.01594

AUTHOR INDEX

SUBJECT INDEX

A

B

E

I

CRC PUBLICATIONS OF RELATED INTEREST

CRC HANDBOOK OF CHEMISTRY AND PHYSICS, 60th Edition
Edited by **Robert C. Weast, Ph.D.**, Consolidated Natural Gas Co., Inc.
This Handbook is the definitive reference for chemistry and physics and maintains the tradition that has earned it the reputation as the best scientific reference in the world.

CRC HANDBOOK OF ENVIRONMENTAL CONTROL
Edited by **Richard G. Bond, M.S., M.P.H.**, and **Conrad Straub, Ph.D.**, both with the University of Minnesota.
Designed in a logical sequence to deal with the major questions on environmental control, the Handbook deals with many aspects of the environment with an emphasis on data rather than discussion.

CRC HANDBOOK SERIES IN MARINE SCIENCE
Edited by **J. Robert Moore**, Marine Science Institute, The University of Texas.
A multi-volume series that eventually will explore virtually every aspect of the marine environment, this Handbook presently deals with oceanography, marine products, mariculture, and fisheries.

CRC HANDBOOK OF MATERIALS SCIENCE
Edited by **Charles T. Lynch, Ph.D.**, Wright-Patterson AFB, Ohio.
Providing a current, readily accessible guide to the physical properties of solid state and structural materials, this Handbook is interdisciplinary in approach and content.

CRC HANDBOOK OF RADIOACTIVE NUCLIDES
Edited by **Yen Wang, M.D., D.Sc. (Med.)**, Homestead Hospital, Pittsburgh.
Information compiled from current scientific journals and other sources of scientific data were selected for this Handbook, providing a large fund of information for workers using radioactive nuclides.

CRC HANDBOOK OF TABLES FOR APPLIED ENGINEERING SCIENCE, 2nd Edition
Edited by **Ray E. Bolz, D.Eng.**, Worcester Polytechnic Institute, and **George L. Tuve, Sc.D.**, formerly with Case Institute of Technology.
Covering many fields of modern engineering, this comprehensive reference is designed to serve both the practicing engineer and the engineering student.

INDUSTRIAL CONTROL EQUIPMENT FOR GASEOUS POLLUTANTS
By **Louis Theodore, D.Eng., Sc.**, Manhattan College, and **Anthony Bounicore, M.Ch.E.**, Entoleter, Inc.
Directed toward the fundamentals and design principles of industrial control equipment for gaseous pollutants, this reference includes pertinent data with appropriate practical applications.

RECENT DEVELOPMENTS IN SEPARATION SCIENCE
Edited by **Norman N. Li, Sc.D.**, Exxon Research and Engineering Co.
Intended for both postgraduates and professionals, this multi-volume reference set discusses recent developments in the science and technology of separation and purification.

TRACE ELEMENT MEASUREMENTS AT THE COAL-FIRED STEAM PLANT
By **W. S. Lyon, Jr., B.S., M.S.**, Oak Ridge National Laboratory.
This is an examination of an exhaustive trace element balance made at the T. A. Allen Steam Plant in Memphis, Tennessee by an interdisciplinary group of scientists and technologists.

CRC CRITICAL REVIEWS IN ANALYTICAL CHEMISTRY
Edited by **Bruce H. Campbell, Ph.D.**, J. T. Baker Chemical Co.

CRC CRITICAL REVIEWS IN ENVIRONMENTAL CONTROL
Edited by **Conrad P. Straub, Ph.D.**, University of Minnesota.

CRC CRITICAL REVIEWS IN SOLID STATE SCIENCES
Edited by **Donald E. Schuele, Ph.D.**, and **Richard W. Hoffman, Ph.D.**, Case Western Reserve University.

EVALUATED KINETIC DATA FOR HIGH TEMPERATURE REACTIONS
By **D. L. Baulch, M.Sc., Ph.D., D. D., Drysdale, B.Sc., Ph.D.,** and **D. G. Horne, B.Sc., Ph.D.,** University of Leeds, England. This two-volume reference selects specific homogenous gas phase reactions of importance in high temperature systems and offers a comprehensive tabulation of the available reaction rate data with a critical evaluation of the existing data.

EVALUATED KINETIC DATA ON GAS PHASE ADDITION REACTIONS
By **J. A. Kerr, B.Sc., Ph.D., D.Sc.,** and **M. J. Parsonage, A.R.I.C., Ph.D.,** University of Birmingham, England.
Quantitative kinetic data on gas-phase addition reactions are offered in this volume, presented in a logical and consistent manner with a critical assessment of their validity based on presently accepted methods of kinetic analysis.

FUNDAMENTAL MEASURES AND CONSTANTS FOR SCIENCE AND TECHNOLOGY
By **Frederick D. Rossini,** Rice University.
Invaluable to working scientists, as well as students in science or engineering, who need to know the basis and current status of the measurements involved in their respective disciplines.

CRC FORUM ON ENERGY
Editor-in-Chief, Robert J. Budnitz, Ph.D., University of California at Berkeley.
The Forum consists of four sessions with each session featuring four one-hour tapes and research papers exploring technical and energy policy issues with emphasis on societal concerns.

Please forward inquiries to CRC Press, Inc., 2000 N. W. 24th Street, Boca Raton, Florida 33431.